▶ Infantry Combat Medics in Europe, 1944–45

Other Palgrave Pivot titles

Asoka Bandarage: Sustainability and Well-Being: The Middle Path to Environment, Society, and the Economy

Panos Mourdoukoutas: Intelligent Investing in Irrational Markets

Jane Wong Yeang Chui: Affirming the Absurd in Harold Pinter

Carol L. Sherman: Reading Olympe de Gouges

Elana Wilson Rowe: Russian Climate Politics: When Science Meets Policy

Joe Atikian: Industrial Shift: The Structure of the New World Economy

Tore Bjørgo: Strategies for Preventing Terrorism

Kevin J. Burke, Brian S. Collier and Maria K. McKenna: College Student Voices on Educational Reform: Challenging and Changing Conversations

Raphael Sassower: Digital Exposure: Postmodern Postcapitalism

Peter Taylor-Gooby: The Double Crisis of the Welfare State and What We Can Do About It

Jeffrey Meyers: Remembering Iris Murdoch: Letter and Interviews

Grace Ji-Sun Kim: Colonialism, *Han*, and the Transformative Spirit

Rodanthi Tzanelli: Olympic Ceremonialism and the Performance of National Character: From London 2012 to Rio 2016

Marvin L Astrada and Félix E. Martín: Russia and Latin America: From Nation-State to Society of States

Ramin Jahanbegloo: Democracy in Iran

Mark Chou: Theorising Democide: Why and How Democracies Fail

David Levine: Pathology of the Capitalist Spirit: An Essay on Greed, Hope, and Loss

G. Douglas Atkins: Alexander Pope's Catholic Vision: "Slave to No Sect"

Frank Furedi: Moral Crusades in an Age of Mistrust: The Jimmy Savile Scandal

Edward J. Carvalho: Puerto Rico Is in the Heart: Emigration, Labor, and Politics in the Life and Work of Frank Espada

Peter Taylor-Gooby: The Double Crisis of the Welfare State and What We Can Do About It

Clayton D. Drinko: Theatrical Improvisation, Consciousness, and Cognition

Robert T. Tally Jr.: Utopia in the Age of Globalization: Space, Representation, and the World System

Benno Torgler and Marco Piatti: A Century of *American Economic Review*: Insights on Critical Factors in Journal Publishing

Asha Sen: Postcolonial Yearning: Reshaping Spiritual and Secular Discourses in Contemporary Literature

Maria-Ionela Neagu: Decoding Political Discourse: Conceptual Metaphors and Argumentation

Ralf Emmers: Resource Management and Contested Territories in East Asia

Peter Conn: Adoption: A Brief Social and Cultural History

Niranjan Ramakrishnan: Reading Gandhi in the Twenty-First Century

Joel Gwynne: Erotic Memoirs and Postfeminism: The Politics of Pleasure

palgrave▶pivot

Infantry Combat Medics in Europe, 1944–45

Tracy Shilcutt
Associate Professor of History, Abilene Christian University

DOI: 10.1057/9781137347695

© Tracy Shilcutt 2013

All rights reserved. No reproduction, copy or transmission of this publication may be made without written permission.

No portion of this publication may be reproduced, copied or transmitted save with written permission or in accordance with the provisions of the Copyright, Designs and Patents Act 1988, or under the terms of any licence permitting limited copying issued by the Copyright Licensing Agency, Saffron House, 6–10 Kirby Street, London EC1N 8TS.

Any person who does any unauthorized act in relation to this publication may be liable to criminal prosecution and civil claims for damages.

The author has asserted her right to be identified as the author of this work in accordance with the Copyright, Designs and Patents Act 1988.

First published 2013 by
PALGRAVE MACMILLAN

Palgrave Macmillan in the UK is an imprint of Macmillan Publishers Limited, registered in England, company number 785998, of Houndmills, Basingstoke, Hampshire RG21 6XS.

Palgrave Macmillan in the US is a division of St Martin's Press LLC, 175 Fifth Avenue, New York, NY 10010.

Palgrave Macmillan is the global academic imprint of the above companies and has companies and representatives throughout the world.

Palgrave® and Macmillan® are registered trademarks in the United States, the United Kingdom, Europe and other countries

ISBN: 978–1–137–34770–1 EPUB
ISBN: 978–1–137–34769–5 PDF
ISBN: 978–1–137–34768–8 Hardback

This book is printed on paper suitable for recycling and made from fully managed and sustained forest sources. Logging, pulping and manufacturing processes are expected to conform to the environmental regulations of the country of origin.

A catalogue record for this book is available from the British Library.

A catalog record for this book is available from the Library of Congress.

www.palgrave.com/pivot

DOI: 10.1057/9781137347695

*For all the combat medics.
Thank you.*

Contents

Acknowledgments	vii
List of Abbreviations	viii
Introduction	1
1 Chalkboard Training	10
2 Baptism of Fire	29
3 Combat Reality	49
4 The Battalion Aid Station	63
5 Day-to-Day Health	82
6 Company Aid Men	104
Conclusion	119
Select Bibliography	125
Index	133

Acknowledgments

It is difficult to find words to express my gratitude to the many dedicated people who supported and assisted me in this project. To every veteran who responded to the questionnaires and corresponded with me I offer my sincere thanks. Words fail to convey the emotional upheaval inherent in a project such as this, and despite that reality these men shared with me their stories.

Thanks to Dr. Mark T. Gilderhus for his faith in me and his untiring support of the project. Thanks also to Dr. John L. Robinson for his commentary and critical review of early drafts. Appreciation is also due to members of the faculty at Abilene Christian University, Dr. Fred Bailey, Dr. Vernon Williams, and Dr. Cole Bennett, and at Texas Christian University, Dr. Kenneth R. Stevens, Dr. Claire Sanders, and Dr. Steven Woodworth. Neel Price, M. D., Kyle Sheets, M. D. and Jim Morrison, M. D. corrected many misconceptions of emergency medical care. Special thanks to Brent Fisher, Jackie Beth Shilcutt, Megan H. Robinson, J. Chase Beakley, and Corey Self for close attention to detail in preparing the manuscript. I am grateful to Michael Doubler for his critique and insight.

The staffs of the National Archives, and the Interlibrary Loan departments at Texas Christian University and Abilene Christian University deserve recognition for their guidance and diligence. A special note of thanks to David Keough of the U.S. Army Center of Military History Institute at Carlisle Barracks, Pennsylvania. This work would not have been possible without the support of Bryan Shilcutt, Jackie Beth Shilcutt, and Erin Shilcutt—thank you.

List of Abbreviations

AEF	American Expeditionary Force
A&P	Ammunition & Pioneer
ASF	Army Service Forces
ASTP	Army Specialized Training Program
BAR	Browning automatic rifle
BAS	battalion aid station
CI	Combat Interviews
CIB	combat infantry badge
CMB	combat medical badge
CP	command post
EC	Eisenhower Center
EMT	Emergency Medical Tags
ETO	European Theater of Operations
FAO	Field Artillery Observer
FAS	forward aid stations
FM	Field Manual
FUO	Fever of Unknown Origin
HI	Hospital Interviews
MAC	Medical Administrative Corps
MAC-OCS	Medical Administrative Corps-Officer Candidate School
MC	Medical Corps
MHI	Military History Institute
MRTC	Medical Replacement Training Center
MTP	Mobilization Training Program
NARA	National Archives and Records Administration
OCS	Officer Candidate School
RG	Record Group

SBSA	School for Battalion Surgeon's Assistants
SIW	self-inflicted wounds
TM	Technical Manual
UAR	Unit Annual Report
VD	veneral disease
WIA	Wounded in Action

palgrave▸pivot

www.palgrave.com/pivot

Introduction

Abstract: *The introduction provides a brief overview of the historical development of the United States Medical Department from 1799 to 1939 with attention to World War I personnel shortages and medical evacuation procedures. The introduction defines key words used in the study on World War II first echelon medics including Battalion Aid Station, "medic," "combat medic," and "aid man." Further, it describes the multi-tiered evacuation system used by the infantry in European campaigns of 1944–45. The introduction provides a note on sources, including questionnaires prepared by the author.*

Shilcutt, Tracy. *Infantry Combat Medics in Europe, 1944–45*, Basingstoke: Palgrave Macmillan, 2013. DOI: 10.1057/9781137347695.

The action lasted less than 30 minutes, but in the early spring of 1945, an "intense firefight" east of the Rhine River drove the American soldiers to ground. As they scuttled for shelter, the "zing and pop of [the] bullets" attended continuous tree bursts, which rained on F Company of the 309th Infantry. Remarkably, the unit suffered only one casualty, Robert Gregory, who was writhing and in great pain. When medic John Collins reached Gregory, he quickly cut the uniform away exposing the wound. There was no blood, "merely raw flesh" shaped like the sideways bullet that entered near the kidneys. Collins realized the random ricochet would prove fatal, so the medic spoke gently to Gregory, telling him that he would be headed home. Gregory died a short time later and Collins surely grasped more clearly his limitations as a company aid man.[1]

Just two years after the war's end, Collin's division published its official history, but it scarcely portrayed the complexity of the aid man's job. The spare description analytically depicts a typical combat medic moving under fire to the side of a wounded man, administering morphine, and then calling for a litter squad; the bearers swiftly remove the wounded soldier to a battalion aid station (BAS) where a knowledgeable doctor gives him professional attention.[2] A subtle tone of admiration underlies the text, but the clinical account implies that front line medics somehow managed the combat environment in some pre-determined formulaic pattern. The depiction conveys nothing of the appalling wounds, the brutally random nature of war, nor the maddening bedlam swirling around field medics at work. In short, while the narrative acknowledges the combat medic as a vital cog in the war machinery, it conveys no understanding of the front line medical soldier, laboring in a surreally violent world of bewildering complexity. The dual nature of combat exposes war's striking irony: the charge to close with and destroy the enemy couples with tasks of salvage, that is, the rescue of the wounded infantrymen who might recover and rejoin the fight.

During World War II, combat medics adapted in battle, overcoming shortcomings in the Army's medical system and gross inadequacies in their own training. They initiated critical treatment for the battle wounded, dramatically increasing chances for survival. As these medical soldiers plunged into combat, they at once found their training had scarcely prepared them for an environment marked by unrelenting carnage as the most consistent truth. In a military structure that demanded conformity, combat medics had to adapt, modify, evade, and disregard standard operating procedures in order to save lives. They dared innovatively to impose some small order on their chaotic environment. Infantry medics contributed substantially to the combat effort in Europe, and although there remains a fundamental cognizance that medics gave the combat wounded some chance for life, scant understanding exists of the identity of combat medics, the role of their training, circumstances under which they operated, and how these medical soldiers fitted into the fighting units. This study then focuses on infantry combat medics in Europe, opening a dialogue for understanding the function, humanity, and singular accomplishments of the front line medical soldier during World War II.

For the purposes of this study, the terms "medic" and "combat medic" refer inclusively to any soldiers who worked out of the first echelon battalion aid stations; this designation consequently applies to Medical Corps (MC) officers, Medical Administrative Corps (MAC) officers, non-commissioned officers, medical and surgical technicians, litter bearers, and company aid men who functioned within the battalion-level medical detachments. The term "aid man," by contrast, refers exclusively to those medical soldiers assigned to rifle or weapons companies; these men, such as Private John Collins, had the daily task of rendering immediate aid to the wounded soldier under fire on the front line. And while the fundamental first aid tasks of front line emergency care did not significantly change from those employed in previous wars, the structure under which World War II medical soldiers operated had revised due to lessons learned from past wars, and by larger cultural influences including the professionalization of and specialized training within the medical community.

Overview: historical development of the Medical Department to World War I

The United States Congress approved an Army Medical Department in 1799, and then in 1818 authorized the permanent position of Surgeon General. But until the Civil War, American military battlefield medical care depended primarily upon the cooperation of regimental surgeons working within a military system that failed to prioritize a policy for casualty evacuation. The Medical Department's failure to consolidate casualty treatment and evacuation during conflicts with frontier Indians and with Mexico meant that American military medical doctrine was in a sorry state when the Civil War began.[3]

Both the Union and Confederate Armies suffered from the lack of a coordinated system to care for and evacuate wounded rearward, so during the early battles of the war, wounded might languish on the field for up to a week. Additionally, Sanitary Commission volunteers rather than Medical Department personnel ultimately provided definitive care for the wounded in the hospitals. Perhaps one of the most significant adaptations employed by both Northern and Southern medical practices was the emergence of evacuation systems that utilized ambulance corps modeled after those in the French Army. This substantially reduced the time the wounded lay unattended.[4]

Following the Civil War the Medical Department largely reduced its numbers as front line military medicine increasingly came under the control of civilian physicians. But, even as United States Army medical doctrine stagnated from 1870 to 1890, scientific discoveries and increased standardization of medical education transformed American medicine and this professionalization had ramifications for military medicine in the last decade of the nineteenth century.[5]

Military surgeons increasingly accepted tools and techniques standard in the larger world of Western medicine, which allowed for a better understanding of the processes of disease.[6] The military mirrored broader cultural trends toward better education for physicians when the Medical Department Army Medical School opened in 1893. Yet, the abysmal performance of the MC during the War with Spain in 1898 resulted in a reorganization of the Medical Department. During the early decades of the twentieth century, Medical Department officers Walter Reed, Leonard Wood, and William Gorgas led the medical world as they made dramatic progress toward controlling tropical diseases.[7]

When the First World War erupted, American volunteers and medical units numbered among the first to aid in Allied casualty care and evacuation efforts. The Medical Department followed the American Red Cross and sent medical military observers to France as early as 1917. Despite this strong initial showing, the Medical Department suffered perpetually throughout the war with shortages of medical personnel and supplies.[8]

When the United States declared war, the Medical Department was targeted for reductions. Planners had proposed that the enlisted force of the Medical Department would have a relative strength of 10 percent of the Army, but within only two months, the Tables of Organization reduced the allowance to a 9 percent ratio.[9] Because the Medical Department made the needs of the front its priority, the shortage crisis became acute when the United States committed combat troops. Further, a lack of medical replacement soldiers compromised care at every level, when the Army diverted medical soldiers intended for front line care to more rearward facilities focused on disease care. Disease was not a new issue; it had dominated past war health concerns, but the shortsighted projections of allowances of medical personnel directly impacted the effectiveness of the front line medical care. Yet, even with severely deficient numbers, front line treatment and evacuation successes were due to the devotion of duty by medical soldiers assisted by the Red Cross, AEF (American Expeditionary Force) band members, unit line soldiers, and impressed German soldiers.[10]

Additionally, front line medical soldiers worked within a confusing evacuation system which left each fighting unit to develop its own priorities for treatment and evacuation of the wounded. With no single method "universally employed" for evacuation, the most consistent units organized their evacuation systems on established British and French methods.[11] In general, these methods allowed for trench medical personnel to remain with companies on the line (company aid posts) and for fixed positions of front line installations (battalion aid stations). On the other hand, the evacuation system for open warfare was more dependent upon mobile battalion aid stations than on company aid posts, as the forward pace of the battle meant that the medical soldiers constantly resituated closer to the front.[12] Lessons learned from World War I regarding a multi-echelon medical evacuation system impacted future training programs for medics, but the systemic disregard by the Army of the Medical Department meant that World War II combat medics faced numerous challenges.

Definitions and overview: World War II combat medics

World War II American soldiers fought within a multi-tiered medical aid structure, under a chain of care designed to provide increasingly definitive medical treatment as the wounded men moved incrementally to the rear. Proximity to the firing line and its dangers defined the care and evacuation system, with primary imperatives focused on immediate treatment of the casualty with minor wounds so that he could return quickly to the fray, or on efforts to stabilize and keep alive the seriously wounded soldier, moving him rearward to better equipped and safer facilities.[13] The company aid man constituted the essential linchpin of this expansive strategy for the combat infantryman.

As the most forward of the first echelon medical soldiers, the aid man technically remained under the control of a BAS commanding officer. In practice, though, he left the comparatively safe confines of the aid station to live with the combat line companies because his responsibilities centered on providing immediate care to the wounded where they fell. After the aid man tended to the casualty, litter bearers carried the wounded soldier to the BAS where teams of medical soldiers worked quickly to treat and then possibly evacuate him. These stations, located anywhere from close on the line to a mile behind the battlefield, ordinarily served as the first stopping point for the wounded at the most rearward of first echelon care.

Operating independently between Battalion Headquarters and the combat line, a battalion surgeon, schooled as a medical doctor, commanded the BAS. These soldiers moved in tandem with their advancing combat unit and set up shop as close to the line as feasible in order to shorten the casualties' travel time. Location depended upon an engagement's environment, and medical men placed aid stations variously in burned out buildings, under tents and shelter halves, in dugouts, beneath hastily placed logs, and even in the back of vehicles. Once the wounded soldier received BAS care, first echelon responsibilities ended and second echelon medical soldiers facilitated movement back through the evacuation system. Thus the linear arrangement of the medical system began with initial treatment by the company aid man, transport by litter bearers, and further care by BAS medics.

The second echelon consisted of a division's organic medical unit, its Medical Battalion, usually bivouacked with Division Headquarters two miles or more behind the field of battle. This level administered care for the soldiers who might promptly recover to fight again, and sent the more critically wounded back to the third echelon, the level furthest removed from the battlefield. Located well away from the combat zone, the third echelon included facilities such as general hospitals and rehabilitation centers and provided a sterile environment for the work of doctors, nurses, and medical technicians. Here, seriously wounded men had the chance to recuperate in a fire-free area or to be sent home for extended treatment.

Sources

The body of historical literature devoted to military medicine during World War II focuses on the corporate success of the Medical Department, giving little or no attention to the combat medic. Historians appropriately credit medical advances, penicillin, sulfa drugs, and blood transfusions, and the structural framework of the evacuation system for the impressive survival rate of combat wounded. Yet every success depended fundamentally upon those front line medics who initially tended battlefield wounded so that they might move back to rear echelon facilities offering more sophisticated treatment. At the same time, unit histories laud in passing the medical soldiers' efforts, but these narratives unsurprisingly focus on the combat team's tactics and strategies rather than on the medics' distinct and unique contributions to that team.

A recent trend in military history involves efforts to refocus attention on the citizen-soldiers who worked together, learning on the battlefield while prosecuting the war successfully.[14] Although many of these studies briefly acknowledge the assistance of the combat medic, they neglect the role of the medical soldiers. Indeed, the materials rely on a severely limited number of medic experiences, those whose stories appear in published memoirs or letters.[15] Holdings at the National Archives and the Military History Institute (MHI) provide extensive institutional records, yet they lack depth concerning first echelon medics because few interviews conducted during the war included medics as participants. At the same time, regimental and battalion records peripherally acknowledge medics as courageous men committed to their tasks, but include only a handful of personal accounts from the medical soldiers themselves. For these reasons, the sources undergirding this study begin with questionnaires designed to address the gap of information concerning the generally unexamined role of the infantry combat medic in Europe, 1944–45.

Questionnaires facilitated access to a population of veterans with combat experience in Europe, either as first echelon medics, or as infantrymen.[16] Their participation signals a willingness to reveal long-shrouded memories as they approach the later years of their lives. Former medics who responded to the questionnaires often confessed that they had carried their stories untold for more than half a century, yet intense memories remained vivid long after some had forgotten the names of the places where they fought. As the veterans recalled events, long-buried emotions resurfaced, compelling some to put aside temporarily the questionnaires; for others, the experience simply proved too intense and they could not complete the task.

Each soldier experienced the war uniquely, and each retained highly personal impressions of the conflict. Yet, patterns developed in their stories, emerging to give medics a fresh and unique voice in combat's narrative. Significantly, their stories reveal that the Army's grossly inadequate training shaped a naïve perception of the combat medic's role. Despite the lack of realistic preparation and the horrific circumstances in which they worked, combat medics proved extraordinarily capable, creative, and committed to doing anything necessary to perform their duties as the essential first link in the wounded soldier's life chain. Their combat competencies and experiences were born in blood.

Notes

1. Frank J. Irgang to author, (n.d.); 78th Infantry Division Veterans Association, *The Flash* (January 2001): 95.
2. United States, Army, 78th Division (1947) *Lightning: The Story of the 78th Infantry Division* (Washington, DC: Infantry Journal Press), p. 106.
3. R. V. N. Ginn (1997) *The History of the US Army Medical Service Corps* (Washington, DC: Infantry Journal Press), pp. 3–8.
4. Ginn, *History*, pp. 6–8.
5. Ibid., pp. 9–19; J. H. Cassedy (1991) *Medicine in America: A Short History* (Baltimore: The Johns Hopkins University Press), pp. 88–91; and P. Starr (1984) *The Social Transformation of American Medicine: The Rise of a Sovereign Profession and the Making of a Vast Industry* (New York: Basic Books), pp. 114–126.
6. Cassedy, *Medicine*, pp. 88–91; and Starr, *Social Transformation*, pp. 114–126.
7. Cassedy, *Medicine*, pp. 114.
8. United States (1991) *United States Army in the World War, 1917–1919*, Volume 15 (Washington, DC: Center of Military History), pp. 370–371; and C. Lynch, J. Ford, and F. W. Weed (1925) *The Medical Department of the United States Army in the World War*, Volume VIII (Washington: GPO).
9. Lynch, Ford, and Weed, *Medical Department*, Volume VIII, pp. 13–15; United States (1989) *United States Army in the World War*, Volume 4 (Washington DC: Center of Military History), p. 530.

10 Lynch, Ford, and Weed, *Medical Department,* Volume VIII, pp. 18–28.
11 Ibid., "universally employed," p. 105 and pp. 90–92, 105–113.
12 Lynch, Ford, and Weed, *Medical Department,* Volume VIII, pp. 91–92, 105–130.
13 Regimental aid stations served primarily as administrative and supply units to the BAS, locating with regimental headquarters. Although individual units functioned as first aid stations as needed, their fundamental activities remain outside the scope of the study except as the personnel directly interacted the BAS medics.
14 See S. Ambrose (1997) *Citizen Soldier: The US Army from the Normandy Beaches to the Bulge to the Surrender of Germany* (New York: Simon and Schuster); J. C. McManus (1998) *The Deadly Brotherhood: The American Combat Soldier in World War II* (Navato, CA: Presidio); G. F. Linderman, *The World Within War: America's Combat Experience in World War II* (Cambridge: Harvard University Press, 1997); M. D. Doubler (1994) *Closing with the Enemy: How GIs Fought the War in Europe* (Lawrence: University of Kansas Press).
15 R. Bradley (1970) *Aid Man!* (New York: Robert Bradley); F. J. Irgang (1949) *Etched in Purple* (Caldwell, OH: The Caxton Printers); William Shinji Tsuchida (1947) *Wear It Proudly: Letters* (Berkeley: University of California Press).
16 Questionnaires facilitated access to a population of veterans with combat experience in Europe, either as first echelon medics, or as infantrymen. This study utilizes 57 questionnaires prepared by the author: 53 former combat medics and 25 former infantrymen returned completed questionnaires and granted permission for use. Hereafter all questionnaires are referred to as "Questionnaire."

1
Chalkboard Training

Abstract: *This chapter highlights the stateside training experiences for those who would become front line medics, 1944–45. Lessons learned from World War I should have alerted World War II planners to possible medical and evacuation problems, but medical soldier training regimens from 1920 to 39 reveal that the Army continued to marginalize Medical Department combat preparedness. Instead, the Army designed training programs to forge a standardized medical soldier who could be plugged in anywhere along the chain of evacuation. The unimaginative training short-changed the battlefield medics.*

Shilcutt, Tracy. *Infantry Combat Medics in Europe, 1944–45*, Basingstoke: Palgrave Macmillan, 2013. DOI: 10.1057/9781137347695.

On Sunday 7 December 1941 18-year-old Paul Winson sat stunned in New York's Empress Theater as its employees interrupted the film with news of the Japanese attack on Pearl Harbor. The youngest of five children, Winson went the next day to enlist in the Navy, but diminished vision in his right eye disqualified him for service. By 1943, however, "the military wasn't so particular," and the Army drafted him, judging his sight sufficient for medical duties. Marked for "limited service," Winson trekked cross-country to the Medical Replacement Training Center (MRTC) at Camp Grant, Illinois. Together with a core group of 120 fellow New Yorkers he plunged into "intensive" basic infantry training, without weapons. More importantly, as prospective medical soldiers they received no "adequate" first aid instruction.[1]

Despite the global conflict then raging, "tranquility" rather than a "climate of death" permeated the camp, leaving Winson with a "light-hearted attitude" about training and his looming tasks. Army instructors carried out much of the schooling in classroom settings, while periodic first aid sessions held in the wooded terrain emphasized locating and moving simulated casualties to an aid station. These poorly constructed training exercises, intended to introduce the real dangers of combat, inevitably deteriorated as teenagers found humor in even the most serious lessons. The medical trainees role-played litter bearers and casualties, and these scenarios often turned "comical" as the men tried to hoist and then haul the litter patients. So in an effort to accomplish their tasks more easily, his fellow trainees assigned Winson, one of the "lightest and smallest" of the group, the recurring role of wounded soldier.[2] This light-hearted approach defined Winson's experiences at Grant.

During a hands-on seminar on poison gas, MRTC instructors stressed the probability of chemical warfare, drilling the New York recruits on the identification of various gases. A "Missouri farm boy" sergeant, leading the discussion on phosgene, assured the group that they would have no trouble recognizing this chemical because it smelled like "new mown hay." The fledgling soldiers looked doubtfully at one another, each "with raised eyebrows," because not one of the city boys had any idea what any sort of hay smelled like. When they tried to explain to the sergeant the peculiar odors of the New York City subway, each recognized that the other had no frame of reference. The seriousness of chemical warfare vanished in laughter, all too typical of a training regime remote from the shock of combat.[3]

Following basic training, Winson's indifferent combat training continued at Lawson General Hospital in Atlanta, Georgia, a "topnotch facility," but one focusing on sterile procedures rather than the emergency treatment that the unsanitary front lines demanded. While at Lawson, Winson assisted in the operating room, managed surgical equipment, and completed course work in anatomy and pharmaceutical procedures, training ill-suited for a nascent front line medic. Following his Lawson stint Winson moved on to Camp Reynolds, Pennsylvania, where in the days just prior to his departure for the European Theater of Operations (ETO) he cleaned pot-bellied stoves and picked up cigarette butts on the post.[4]

Winson's training suggests the Army envisioned him serving in a rear echelon medical environment, but wars rarely conform to expectations. When Winson arrived in the ETO, the Army quickly plugged him into the 30th Infantry Division as a replacement company aid man, and he found that his training had ill-prepared him for the realities of a combat medic's duties. There were no autoclaves or boiling water on the battlefield, no sterile procedures in the battlefield muck.[5]

Winson's introduction to war is certainly not a new story. The battlefield has always been a hard teacher and by World War II, general Army doctrine recognized that on-the-job training was to be expected. But with the Army's training directives and materiel focused on teaching soldiers to kill, combat medics participated in training programs that relied on chalkboard explanations for initial aid and the evacuation system. The horrific realities of war soon made it plain that the line of battle that had been chalked on training boards was far more fluid than trainees had been led to believe. Successful medics swiftly adapted to combat conditions, insuring their own survival as well as that of their wounded comrades. Learning on the fly and under deadly fire, they discarded or radically modified prescribed medical techniques, discovered ways to utilize the changing terrains for their own protection and that of the wounded, and coped with unanticipated long-term problems. Yet lessons taught by World War I should have convinced the Army of the crucial role of front line medic and the centrality of realistic training.

American medical soldiers had been among the first to see combat during World War I in support of British and French troops. These first-wave soldiers worked within an established evacuation model, which included staged treatment that began in the trenches and continued as the wounded moved rearward by motorized transport.[6] Stateside, the United States mobilized for the Great War, with medical soldiers participating in four-week training programs emphasizing personal hygiene, sanitary service, and some field training. American planners read reports from France concerning combat medical demands, and they requested increased strength for the Medical Department, stressing that the "needs of the front had precedence." But Tables of Organization allowances shorted medical personnel, allocating 2 percent less than had been authorized.[7] While military medicine is not generally a compassionate endeavor, this culture of inattention to the wounded soldier at the front would persist through World War II.

Events in France only compounded this organizational shortage. For one, medics intended to provide emergency first aid for the wounded at the front were diverted to care for diseased soldiers in the rear. Not only was care for the combat wounded on the line sacrificed for care of sickly troops, the AEF reassigned at least 1,000 medical soldiers to the front line as combat troops.[8] The resultant "acute shortage" of front line medical personnel pressed other soldiers, including band members and German prisoners, into service hauling litters. The Army's chief surgeon judged the situation as near disastrous, suggesting that only devotion to duty and outside assistance saved the AEF.[9] Analysis revealed that larger numbers of better-trained medical soldiers must be stationed at the front.

Along with a shortage of medical soldiers, Americans in World War I lacked a systematized evacuation structure. While elements of the American troops had worked with and learned from the British and French, the AEF failed to institutionalize a consistent organizational model. Some units did successfully adapt to meet the challenge of front line medical care. In general, effective units established a company aid post forward of the BAS, staffing it with one battalion surgeon and two enlisted men. These medical soldiers treated the wounded where they fell and readied them for litter transport to the trench station.[10] Additionally, in these static positions an assortment of trench diseases such as trench-foot, trench-mouth, trench-nephritis, and trench fever complicated the job of the front line medical soldiers even though Americans saw limited trench action.[11]

Combat medical soldiers caring for units that moved from trench to open warfare had to adapt quickly as BAS moved out of the trenches and onto or near the battle lines. The "Manual for the Medical Department" proved less than prescriptive, and each fighting unit developed its own priorities for treatment and evacuation with no single method universally employed, but in each case, the mobility of the BAS allowed it to re-locate repeatedly to be as close to the line of action as possible.[12] Beyond the care for the wounded, medical soldiers also faced a challenge in differentiating war neuroses from shell fright, mental and physical fatigue, or malingering. Additionally, these cases "seriously complicated the problems of evacuation."[13] The post-war report of General John J. Pershing labeled the evacuation of the sick and wounded as a "difficult" problem during the battle period.[14] Other analyses noted that a number of officers of "high rank were convinced" that the Medical Department should not be advised in advance of impending combat activities. The report went on to note an overall failure by the AEF to consult Medical Department personnel appropriately and that for the future "military objects can be attained only considering the military machine as composed of numerous reciprocating parts, each striving towards a common end."[15] This systemic marginalization of the Medical Department in World War I should have been a lesson learned, but the hierarchy issues surfaced again in World War II.

Medical training regimens for those who would serve as ETO combat medical personnel commonly fell into three time frames: peacetime training (1920–39), mobilization training (1939 to December 1941), and wartime training (1942–45). During each period the Army's intense focus on creating citizen-soldiers shaped the training programs. At the end of World War I an isolationist Congress reduced the Medical Department from ten percent of the total Army personnel to five percent. During the inter-war years this paucity of medical personnel engendered a perception that the medical soldier's value lay primarily as a provider of routine, peacetime health care. The Army consequently relegated medically oriented field training to secondary concerns. Peacetime training programs for medical soldiers, as a result, neglected the most basic applied tactical medical operations in favor of hit-and-miss on-the-job training in professional facilities for enlisted men and officers alike. From 1920 to 1939 the Army and its Medical Department failed to emphasize the most fundamental component of medical combat function: field exercises that highlighted first aid and evacuation.[16]

In 1939 the deepening international crisis forced mobilization on the United States. Increasing numbers of soldiers flooded into training centers, straining the already under-strength Medical Department and further jeopardizing the field readiness of medical soldiers. Although the Medical Department had increased to 7.4 percent of total Army strength by the time of Pearl Harbor, the overriding demand to administer routine medical examinations and care to incoming soldiers limited enlisted medical basic training to brief introductions to military discipline and the fundamentals of soldiering together with occasional field exercises. Once again the Army marginalized the field readiness for those medical soldiers destined for combat. Training continued to neglect tactical medical operations in favor of programs focusing on peacetime tasks such as physical check-ups and the nuances of sick call rather than on emergency trauma treatment and movement of the wounded rearward.[17]

Even as the United States prepared for combat in Europe the Medical Department continued to rely on chalkboard explanations of evacuation processes rather than incorporating additional field training to emphasize lessons learned from the recent North Africa theater. Rather than school medics for the cataclysmic horrors of the battleground, the Army, which demanded an average of 15 weeks of training in clerical tasks for medical administrators, could at the same time casually dragoon a cook into the more critical and demanding role of the combat medic.[18]

Most enlisted medical trainees received formal instruction at MRTC. Three MRTCs functioned as the primary training locales for enlisted men who underwent a medical basic program during the mobilization period: Camp Lee, Virginia (activated January 1941, moved to Camp Pickett, Virginia, June 1942); Camp Grant, Illinois (March 1941); and Camp Barkeley, Texas (July 1941). Although the Army purposed to shape a single enlisted basic medical training program, distinctive weather and geographic conditions combined with varying levels of competency of the training personnel and systemic indiscipline to render each camp training experience unique. In addition, the Medical Department prepared their own support personnel including cooks, mechanics, chauffeurs, buglers, and the like. All this led to the extraordinary reality that although technically part of the Medical Department, many of those trained at the MRTCs received little or no medical training.[19]

Upon arrival at an MRTC, a typical recruit shuffled into a camp theater for orientation, listening as the chaplain offered welcome through a "cheering word." Within moments, however, the cheerful climate became but a memory as the executive officer explained Army expectations and demands. Trainees then marched to their barracks and settled into the routine of the Army's strict discipline, designed to transform civilians into soldiers.[20] Medical soldiers at Camp Grant and Camp Lee, the first facilities opened, received 13 weeks of infantry basic training. When the third center, Camp Barkeley, came on stream in late 1941, the MRTC cut schedules at all three facilities to 11 weeks. Even with the reduced training time, the MRTCs' mission remained the same: to prepare medical soldiers for any medical duty required along the chain stretching from battlefield to Zone of the Interior. Although the MRTCs imagined an automaton product, the training programs failed ultimately to produce this hoped-for interchangeable medical soldier. More importantly, beyond this failure to regularize preparation, the Army inexplicably neglected the rudimentary elements of training especially for those who would become front line combat medics.[21]

Shortages of equipment, supplies, and experienced instructors during the mobilization period meant that MRTCs could not always adequately clothe, feed, house, or conduct classes for the inductees. The Mobilization Training Program, designed to introduce the civilian to Army life, emphasized such matters as military courtesy, close order drills, convoy movement, and chemical warfare. The medical soldier-without-arms also received instruction on the organization and function of the Medical Department as related to the Army as a whole. In addition, medical soldiers learned the finer points of guard duty. Other lessons, all too brief, at least enumerated the duties of the company aid man, the BAS personnel, and the role of the BAS.[22]

After the United States entered the war, the Army opened a fourth MRTC at Camp Robinson in Arkansas and further reduced the medical basic program to eight weeks. Over the remaining course of the war, the curriculum and length of the programs varied as training personnel haltingly and inadequately attempted with little success to incorporate lessons learned in combat. The continuing blackboard-bound style of teaching combined with mind-numbing lectures by many of the instructors resulted in unrealistic preparation for tending the horrendous wounds common on the battlefield.[23]

The MRTC ineffectually instructed medical recruits by using puppet shows to "bring home" lessons in military courtesy; simulating casualties, doctors, and nurses, with store-window mannequins; and painting plaster casts to highlight and explain injuries.[24] Private John Sanner, a trainee at Camp Grant during the summer of 1943, learned about medical emergency treatments, hygiene, and "general health information" through lectures and training films. But most explicit in his memory, in common with other young recruits, were the "sobering and revolting" medical training films that dealt with venereal diseases.[25] Captain Frank Miller, a physician who served as a battalion surgeon for the 413th

Infantry Regiment, called the films' graphic nature the "best possible way" to warn the "unwary" of the complications of sexually transmitted diseases.[26]

Yet the MRTCs utilized, unfortunately, no similar techniques to prepare medics for the cataclysmic horrors of the battleground, and a medic's vivid recollection of the ravages of venereal diseases did nothing for the infantryman with a shell-mangled leg. The Army puzzlingly ignored the combat soldier's welfare by failing to utilize extant training methodology. Having attained signal success in its collaboration with such entities as Hollywood's Walt Disney group in producing imaginatively effective training films for a variety of application, the military proved fatally negligent in its concerns for realistically training medics for battlefield casualties.

Enlisted soldiers who completed training at the replacement centers left the camps with ambivalent emotions. They had certainly learned soldiering and many had developed close ties to their training units, but most of these recruit classes broke up as the individuals moved to technical or surgical schools, or were assigned to newly activated organizations. Those who received specialized technical or surgical training gained proficiency in hospital ward procedures rather than combat medicine, while others who transferred to operational units might receive some specialized direction depending upon his particular function within the medical unit: aid man, litter bearer, technician, cook, driver, or clerk.[27]

As the war progressed, training intervals at the MRTCs continued fluctuating between eight and seventeen weeks depending on the need for medical troops.[28] Even though the Army adapted to include additional training specifically tailored for the aid man, the scope of instruction continued to focus on institutional techniques. In addition, the shortened training periods frequently sacrificed the already limited field experiences, so that by late 1944 and early 1945 the "urgent need" for fighting forces in the ETO forced medics into combat even more grossly under-prepared than before.[29]

A combat medical soldier who arrived in the ETO without adequate stateside training found that practical application of his theoretical medical knowledge proved often worthless for his front line unit; indeed many had not expected to be assigned to regimental combat teams as BAS personnel, and "the majority of replacements" had received no unit training of any type.[30] At war's end the Army acknowledged the slipshod nature of the training afforded to these soldiers even while praising the aid men as "important individual[s who were] held in the highest esteem by the combat soldier."[31]

In contrast to those trained at the MRTCs, the Army incorporated more field experiences for those medics who trained stateside with infantry units. One former medic recalled that during stateside training, the medical unit operated as a team, with a clear understanding of "duties and obligations."[32] But the systemic disrespect for the role of the combat medic, which often surfaced during maneuvers, meant that some combat team commanders neglected or ignored the medical teams during training exercises. One 31st Infantry Division combat team commander typified this attitude when he failed to pass on information concerning unit movement to his medical support team while on maneuvers in Louisiana in 1943, leaving them without provisions in bivouac. In this same exercise, Army referees ended the drills once the combat troops made "solid contact," which meant the front line medical troops treated no simulated casualties, focusing instead on minor foot injuries and evacuating soldiers suffering from chigger bites. Even when on maneuvers, BAS personnel rarely had opportunities to establish and work through problems associated with the chain of evacuation and other tasks. One simple, yet profound, result of this failure to work through the field problems would be that once in combat, aid men who had trained with their units on maneuvers would recognize the limitations of their aid kit supplies. Nursing sore feet was a far cry from treating the soldier who had his feet blown off by schu mines.[33]

But stateside, medical trainees such as 18-year-old Robert Reed had no frame of reference for their training experiences so naively judged maneuvers to be "just like battle." Reed's youthful zeal for the field exercises with the 311th Infantry training at Camp Butner, Texas in 1943 sprang from the fact that he spent much of his time on maneuvers scavenging for food. On one three-day field drill, Reed, a company aid man, traveled with the 3d BAS personnel as they followed the infantry into the simulated battle, but once they set up the aid station, he and a fellow field medic left the mock exercise to search for watermelons and water. After a three-mile hike the two medics located a farmhouse where the family eagerly greeted them, treating the young men to a ham dinner with "lots of stuff."

The trainees left the house with full bellies, but rather than returning to the war exercises they continued wandering and eventually located a country store where they bought cookies, cakes, and ice cream. Finally rejoining the field exercise, the medics spent the next two days with their unit in retreat. One consolation for Reed: the weather was "marvellous."[34] While Reed's experience may not have been representative for all aid men, it is indicative of the Army's casual attitude toward the training of front line medical soldiers.

Other BAS enlisted personnel who trained with their infantry combat units, including litter bearers, medical technicians, and surgical technicians, ordinarily assisted in hospital settings rather than practicing for the front line treatment and evacuation. Even so, some who served as combat medics judged that the field exercises taught them how to survive in the open and how to "live with others that smelled worse than you did," if not how to treat wounded comrades.[35]

Certainly their ability to adapt to bivouac circumstances foreshadowed their ingenuity on the battlefield, but the lack of oversight by superiors in these field exercises, coupled with trifling opportunities to hone their combat medical skills, points up the Army's failure to grasp the obvious grave importance of combat medical care. And while there were rare occasions when aid men treated traumatic wounds during the training period, these incidents often occurred on the rifle range rather than in the field, so the evacuation system was once again by-passed in favor of immediate hospital care. Ultimately enlisted men destined to serve as front line medics never grew conversant with the realities that they would face under fire.[36]

Beyond this, the Army sometimes stripped entire enlisted units for deployment, or assigned officers to the BAS at the last minute prior to deployment, so station medics made their adjustments to new leadership on a trial-and-error basis during actual combat operations. Even the Army's post-war analysis judged this as less than desirable, suggesting that the combat units needed to train together a minimum of six weeks for maximum effectiveness.[37]

In contrast to the Medical Department's enlisted men, the officers of the MAC who served as assistants to the battalion surgeons in the BAS benefited from more consistent and thorough tactical operations training. Yet as one MAC officer noted, he and his fellow MAC officers were "entering uncharted territory [and] I am not sure [that] when I was first assigned as an assistant battalion surgeon [the Medical Department] knew what I was supposed to do. I sure as hell didn't have any medical training as such." He was not alone because formal medical schooling formed no part of the MAC programs.[38]

Following World War I the Army had appointed and commissioned MAC officers out of the ranks of enlisted men who had served for a minimum of five years. These officers attended the basic course for officers at the Medical Field Service School in Carlisle Barracks, Pennsylvania that emphasized soldiering, but as with the enlisted men, MAC officers rarely participated in field training experiences. On occasions officer candidates watched demonstration units work through mock evacuation problems, but during the inter-war years the Medical Department assumed that MAC officer trainees would gain needed skills through on-the-job training.[39]

During the mobilization period prior to Pearl Harbor, the Army only slowly established officer candidate schools, judging that the existing numbers of MAC officers were sufficient for initial training plans. By April 1941, however, the increased demands placed on Medical Replacement Training Centers mandated the Army training of additional officers as instructors. The Medical Department responded by creating a Medical Administrative Corps-Officer Candidate School (MAC-OCS) at Carlisle Barracks. The first class enrolled 100 candidates, who trained for 12 weeks until graduation on 1 July 1941. Scheduled for a three-month cycle, the second class of 200 candidates finished the program by 7 December 1941, while the third class of 250 remained in progress. With the U.S. thrust into war, the need for supplementary officers in the newly activated units forced the opening of a second MAC-OCS school at Camp Barkeley, Texas in May 1942.[40]

As Barkeley's school began, 17-year-old Carl Aschoff was finishing his high school education in Burlington, Iowa. His family, listening with "shock [and] worry" to the news of the attack on Pearl Harbor, knew that Aschoff and his older brother must surely serve. A year later Aschoff enlisted, and the Army called him to active duty in March 1943. Sent to Barkeley's MRTC for basic training, Aschoff trudged along on road marches that started with four-mile jaunts and gradually grew to 25-mile treks. He attended lectures, watched training films, and fired on the rifle range. Encouraged to apply for the MAC-OCS, Aschoff remained at Barkeley for this additional training as the Carlisle school had closed. Barkeley had also implemented a pre-OCS school in an effort to weed out soldiers who might not be suited for leadership and to provide initial officer training for those who would continue, and Aschoff joined his MAC-OCS cadre following this pre-training.[41]

The MAC-OCS program had by this time expanded to 17 weeks with the last week of training conducted on bivouac. Aschoff and his classmates often took brief dinner breaks and returned in the evenings to the classroom for several hours' study each week. And while these officer candidates spent more time in field compass and map exercises than did the enlisted trainees, demonstration teams continued as the mechanism for familiarizing them with the critical medical evacuation system. Aschoff graduated from Barkeley's MAC-OCS school in November 1943, confident that his "MAC training gave [him] the background for making decisions" and for precision work. But nothing in his regimen prepared him for the chaos of combat where a shell burst near a small group, killing one, wounding three others, but Aschoff "didn't get a scratch."[42]

A change in the organizational structure of the BAS in 1944 authorized a redistribution of duties for non-medically trained MAC officers. And in response to this increased demand for MAC officers, Carlisle reopened and Barkeley expanded its OCS. MAC officers now filled positions previously held by medically degreed officers in various tactical units, releasing physicians and dentists for other, professional medical duties. The MAC-OCS schools maintained their high standards of training in examinations, inspections, and leadership skills, yet the field work remained somewhat limited, as demonstration units continued modeling the chain of evacuation, while new training aids included miniature battlefields which added useful visuals to the courses. Five hundred trainees entered the program each month, and soon even more were accommodated by beginning a new contingent every two weeks.[43]

Of marked significance to the operation of the BAS was the School for Battalion Surgeon's Assistants (SBSA) organized at Camp Barkeley in early January 1944. This highly specialized facility trained a limited number of newly designated MAC officers specifically for administrative duty as assistant battalion surgeons, a change from the early war years when two medically trained physicians were assigned to each BAS. From 1941 to 1943, the Medical Department had placed two MC officers, both degreed physicians, in each BAS, designating one officer as the battalion surgeon and the other as the assistant battalion surgeon. But as the war ground on into 1944, the demand for physicians increased, forcing the Army to reassign one medically schooled surgeon from each BAS to other needed duties. In turn, the Army assigned one non-medically trained MAC officer to each BAS, relieving the remaining battalion surgeon of administrative duties. Confusingly, the nomenclature *assistant battalion surgeon* did not change as some MAC officers referred to themselves as the Assistant Battalion Surgeon, but after this adjustment, a BAS ordinarily functioned with one medically trained battalion surgeon (MC officer), one non-medically trained officer (MAC officer), and several enlisted soldiers.

The six-week SBSA prepared specially selected MAC officers to act as assistants in all non-medical phases of the administrative and tactical procedures in the operation of the BAS. After fulfilling the requirements of MAC-OCS, a number of officers received this additional preparation. Yet even with the bonus training, emphasis in SBSA continued to rest on tactics and logistical problems rather than on basic medical skills necessary to rendering first aid.[44]

During the years 1944–45, MAC officers assigned to BAS entered the combat zone adequately prepared for their Army leadership roles. They knew how to give orders, organize medical units, and handle paperwork, but battlefield medicine demanded on-the-job training. MAC officers with no medical background found the medical on-the-job training the most demanding part of their combat career. In turn, physicians who became battalion surgeons faced their greatest challenge in acclimating to the world of military courtesy, procedure, and precision.

Most men who would become battalion surgeons had long since begun their careers as medical doctors prior to the bombing of Pearl Harbor, and so the life they had imagined as peacetime healers changed in an instant. Some physicians who had joined the Army Reserve during the interwar years readily grasped what their futures might hold, and those newly completing their medical schooling or rotation understood immediately that combat likely awaited them. For all, it was only a matter of time before the Army called them to active duty.[45]

Taking their medical expertise for granted, the Army sent these would-be medical soldiers to Pennsylvania's Carlisle Barracks for transition into the military life. Carlisle had long been the training ground for medical officers, and the Army slotted these health care professionals among the Medical Corps, Dental Corps, and Sanitary Corps, focusing most of the six-week instruction on military organization and administration. Men who had previously devoted their efforts to indoor activities such as studying or tending to patients, men who had already established their expertise and status, quickly discovered that the structured environment of the Army rendered them greenhorns again as they used "armload(s) of rolled maps" to negotiate the nearby Gettysburg area.[46]

Not surprisingly, these officers found their physical conditioning one of the most difficult disciplines. One physician noted that a single calisthenic exercise left many in his class completely out of breath. Beyond calisthenics, other aspects of the training emphasized physical conditioning through marches and drills. They also learned such "usual aspects of professional courtesy" as saluting and how to wear a uniform.[47]

As the doctors struggled to fuse one professional life to another, many realized that their well-meaning instructors focused on "the gentlemanly traditions of Carlisle," endowing them with "no knowledge at all of combat conditions."[48] One student complained that the instruction "was dreamed up by some easy-chair tactician."[49] To its credit, the Medical Department did instruct the future battalion surgeons on the chain of evacuation, but even this blackboard approach to describing the movement of the wounded to the rear rarely acknowledged the combat line's inevitable fluidity.

Instructors chalked the front line as a straight line at the top of the board with an imagined enemy just beyond, an enemy that courteously failed to violate the chalk line at any point. The explanations for the linkage of the evacuation chain were detailed in equally idealistic, straight line terms, as the battalion surgeons learned that the front troops would remain concealed from the enemy while BAS personnel located their post as close as possible to the line, to a water source, and under some measure of hidden shelter.[50] Company aid men who followed the unit into combat would provide the immediate care for the wounded, and litter bearers would strategically position themselves to initiate the evacuation system. Once the litter bearers removed the wounded to the BAS, the surgeon would then begin his work of triage. Finally, his task as a healer completed in this first echelon of care, and with all of the paperwork neatly done to Army specification, the surgeon would either return the soldier to the line or send him speedily back through the system to the more definitive care facilities.[51]

This theoretical construct ignored the volatile nature of the combat. Instructors did not hint at such scenarios as when a surgeon met a soldier who walked calmly into the BAS and carefully placed his pack and pistol belt in the corner of the room, only then explaining that he had been under a tree that exploded into his legs. The efficient lectures and precise drawings did not foreshadow that when the battalion surgeon cut off the soldier's pants his medical challenge would be to deal with mangled flesh revealing that both calf muscles had been blown away.[52] Certainly some battalion surgeons benefited from additional stateside training in their operational units but others received immediate orders into combat, while still others served at stateside hospitals. 1st Lieutenant Brown McDonald spent two months serving at a stateside infectious disease ward, took courses in tropical diseases, and generally appreciated the "top-notch" facility at which he worked. But as the irony of the Army would have it, McDonald's unit went on line in the middle of 1944's bitter winter in the Vosges Mountains; few tropical ailments challenged him there.[53]

While the Army attempted, if often inadequately, to provide serious and extensive training for its medical officers, the enlisted combat medics who shared the infantry's front line foxholes inexplicably enjoyed no such carefully specialized preparation. Highly praised in combat for their valiant efforts, combat medics' initial training reveals the Army's strange disregard for lessons learned from World War I and even North Africa concerning the critical role these men must inevitably play. Enlisted men trained for an envisioned environment of stability and sterility as the Army introduced them almost casually to the basic concepts of first aid. In training, routine health care activities superseded field maneuvers as they inoculated their fellows, bound up sprained ankles, and dispensed aspirin. In addition, the Army's pragmatic conception of soldiers who could perform medical tasks anywhere within the evacuation system meant that those enlisted men bound for the ETO front lines rarely understood their combat role due to their limited training.

Notes

1 Questionnaire, Paul Winson (31 May 2001); Paul Winson to author, 11 July 2001, in author's possession.
2 All quotes from Questionnaire, Winson; Harry McClain, "A Night's Sleep Ushers Draftee into the Army," *Chicago Daily Tribune* (12 April 1943), p. 2; Unit Annual Report, "Camp Grant, Illinois 1943," Box 220, Record Group 112, National Archives and Records Administration. Depository and reference codes used hereafter are as follows: RG (Record Group); UAR (Unit Annual Report); MTP (Mobilization Training Program); FM (Field Manual); TM (Technical Manual); NARA (National Archives and Records Administration,

3   Questionnaire, Winson; UAR, "Camp Grant, Illinois 1943," RG 112, NARA.
4   Ibid.
5   Ibid.
6   United States (1991) *United States Army in the World War, 1917–19,* Volume 15 (Washington DC: Center of Military History), pp. 370–371.
7   C. Lynch, J. Ford, and F. W. Weed (1925) *The Medical Department of the United States Army in the World War,* Volume VIII (Washington: GPO), pp. 13–19; United States (1989) *United States Army in the World War, 1917–19,* Volume 2 (Washington, DC: Center of Military History) and Volume 4, p. 530.
8   Lynch, Ford, and Weed, *Medical Department,* Volume VIII, pp. 20–22, 90. American combat soldiers arrived in France suffering from measles, mumps, meningitis, and scarlet fever.
9   Lynch, Ford, and Weed, *Medical Department,* Volume VIII, "acute" p. 24 and pp. 23–58, 91–92, and 105–113.
10 Lynch, Ford, and Weed, *Medical Department,* Volume VIII, pp. 66–71, 91–92, 105–131, 360–362; United States, *United States Army, 1917–1919,* Volume 4, p. 328.
11 United States, *United States Army, 1917–1919,* Volume 12, p. 204.
12 Lynch, Ford, and Weed, *Medical Department,* Volume VIII, pp. 105, 130–131 and Volume XV, pp. 1026–1053; United States, *United States Army, 1917–1919,* Volume 4, p. 328.
13 Lynch, Ford, and Weed, *Medical Department,* Volume VIII, p. 65.
14 United States, *United States Army, 1917–1919,* Volume 14, p. 123.
15 Ibid., p. 95.
16 From the end of World War I until mobilization field forces included Regular Army, the National Guard, and the Organized Reserves. The National Defense Act amended in 1920 reduced the Medical Department from wartime strength of ten percent to five percent. R. R. Taylor, W. S. Mullins, R. J. Parks (1974) *Medical Training in World War II* (Washington, DC: Office of the Surgeon General), pp. 1–11; K. R. Greenfield, R. R. Palmer, and B. I. Wiley, (1947, 2004) *The Organization of Ground Combat Troops* (Washington, DC: Center of Military History, United States Army), p. 1.
17 Taylor, Mullins, and Parks, *Medical Training,* pp. 18–23; Greenfield, Palmer and Wiley, *Organization,* pp. 199–203; "Report of the Medical Statistics Division for the Fiscal Year Ending 30 June, 1944," *SGO Annual Reports,* Box 13, RG 112, NARA.
18 Questionnaire, T. William Bossidy (1 July 2001).
19 UAR, "Camp Barkeley 1942–1945," "Camp Robinson 1943," "Camp Grant 1942–1943" and "Camp Lee/Pickett 1942–1943," RG 112, NARA; Taylor, Mullins, and Parks, *Medical Training,* pp. 174–197, 264; R. L. Sanner (1995)

Combat Medic Memoirs: Personal World War II Writings and Pictures (Clemson, SC: Rennas Productions); R. "Doc Joe" Franklin (2006) Medic! How I Fought World War II with Morphine, Sulfa, and Iodine Swabs (Lincoln: University of Nebraska Press), p. 2; Questionnaires, Gerald W. Allen (25 March 2001), Charles C. Cross (10 June 2001), David E. Fought (7 April 2001), Frank J. Irgang (March 2001) and Winson.

20 McClain, Chicago Daily Tribune, p. 2.

21 UAR, "Camp Barkeley 1942–1945," "Camp Robinson 1943," "Camp Grant 1942–1943" and "Camp Lee/Pickett 1942–1943," RG 112, NARA; FM 8–5 (12 January 1942) "Medical Department Units of a Theater of Operations" (Washington, DC: Government Printing Office); FM 8–35 (21 February 1941) "Transportation of the Sick and Wounded" (Washington, DC: Government Printing Office); FM 8–40 (August 1940) "Medical Field Manual. Field Sanitation" (Washington, DC: Government Printing Office); FM 8–45 (October 1940) "Records of Morbidity and Mortality (Sick and Wounded)" (Washington, DC: Government Printing Office); FM 100–10 (December 1940) "Field Service Regulations, Administration" (Washington, DC: Government Printing Office); TM 8–220 (5 March 1941) "War Department Technical Manual: Medical Department Soldier's Handbook" (Washington, DC: Government Printing Office); MTP 8–1 (1941–1944) "Mobilization Training Program 8–1" (Washington, DC: Government Printing Office); Taylor, Mullins, and Parks, Medical Training, pp. 174–197 and 264.

22 MTP 8–1 (January 1941, February 1942).

23 Ibid. UAR, "Camp Barkeley 1942–1945," "Camp Robinson 1943," "Camp Grant 1942–1943" and "Camp Lee/Pickett 1942–1943," RG 112, NARA; Questionnaires, G. Allen, Cross, and Winson; Taylor, Mullins, and Parks, Medical Training, pp. 174–197 and 264. Passive basic instruction through lecture and blackboard instruction served as the primary vehicle of information dissemination for such topics as map reading, convoy movements, anatomy and physiology, medical records keeping, organization of the Army, organization of the Medical Department, personal hygiene, first aid, care of clothing and equipment, identification of hostile combat vehicles, and elements of close order drill.

24 UAR, "Camp Lee/Pickett 1942–1943," RG 112, NARA.

25 Sanner, Memoirs, p. 7.

26 Questionnaire, Frank Miller (April 2001). It appears that some of the training education concerning sexually transmitted diseases correlated with "vice conditions" that threatened the MRTC's host towns; see for instance, Abilene Reporter News, Abilene, Texas (3 April 1941).

27 UAR, "Camp Barkeley 1942–1945," "Camp Robinson 1943," "Camp Grant 1942–1943," and "Camp Lee/Pickett 1942–1943," RG 112, NARA; MTP 8–4 (August 1942).

28 Camp Lee MRTC closed 19 June 1942 and reopened 20 June 1942 at Camp Pickett, Virginia and deactivated October 1943. Camp Grant MRTC officially disbanded 15 October 1944 when the center transferred to Fort Lewis. Camp Barkeley MRTC was designated an ASF (Army Service Forces) Training Center in April 1944 and transferred to Camp Crowder in April 1945. Camp Ellis opened an ASF Training Center in February 1943 and prepared a limited number of medical personnel for combat. UAR, "Camp Barkeley, 1942–1945," "Camp Robinson, 1943," "Camp Grant, 1942–1943," "Camp Lee/ Pickett, 1942–1943," and "Camp Ellis, Illinois, 1944," RG 112, NARA.

29 MTP 8–1 (June 1944); *Report of the General Board, United States Forces, European Theater: Training Status of Medical Units and Medical Department Personnel upon Arrival in the ETO* "Medical Section Study Number 88" (HQ U. S. Forces, 1945), "urgent" p. 4 and pp. 1–8, MHI.

30 *Report of the General Board*, "Study Number 88," "majority" p. 5 and pp. 1–8, MHI.

31 "important" *Report of the General Board*, "Study Number 92" (HQ US Forces, 1945), MHI; Taylor, Mullins, and Parks, *Medical Training*, pp. 174–197 and 264.

32 Questionnaire, William W. Allen (27 November 2001).

33 "Consolidated Report: Surgeon General's Observers for 1942 Maneuvers," Box 217, RG 112, NARA.

34 All quotes from Robert Reed to parents, 21 August 1943, in Reed's possession. Reed trained with the 311th Infantry Regiment; United States, Army, 78th Division (1947) *Lightning: The Story of the 78th Infantry Division* (Washington, DC: Infantry Journal Press), pp. 7–19.

35 *Report of the General Board*, "Study Number 88," MHI; Questionnaires, "live" Les Habegger (2 April 2001), and Wilbur Heinold (2 April 2001), Carroll E. Pomplin (n.d.), Robert R. Reed II (28 April 2001), and Winson.

36 L. Litwak (2001) *The Medic: Life and Death in the Last Days of World War II* (Chapel Hill: Algonquin Books), pp. 25–30.

37 *Report of the General Board*, "Study Number 88," MHI.

38 Questionnaire, "entering" Carl R. Aschoff (14 June 2001); UAR, "Camp Grant 1943," RG 112, NARA.

39 Taylor, Mullins, and Parks, *Medical Training*, pp. 7–13.

40 Ibid., pp. 97–100; UAR, "Camp Barkeley 1942–1945," RG 112, NARA.

41 Questionnaire, Aschoff; UAR, "Camp Barkeley 1942–1945," "Camp Robinson, 1943," "Camp Lee/ Pickett, 1942–1943," and "Camp Grant 1943," RG 112, NARA; Taylor, Mullins, and Parks, *Medical Training*, pp. 110–114. In addition to the Barkeley school, the Army organized additional Prep Schools attached to the MAC-OCS program: one opened in November 1941 at Camp Lee and two others opened in early 1942 at Camps Grant and Robinson.

42 UAR, "Camp Barkeley 1942–1945," RG 112, NARA; Questionnaire, Aschoff.

43 UAR, "Camp Barkeley 1942–1945," RG 112, NARA; Table of Organization and Equipment (TOE), 7 series (1941–1945) (Washington, DC: Government Printing Office); Taylor, Mullins, and Parks, *Medical Training*, pp. 100–114.
44 UAR, "Camp Barkeley 1942–1945," RG 112, NARA; TOE, 7 series; Taylor, Mullins, and Parks, *Medical Training*, pp. 100–114.
45 Questionnaires, Frank R. Ellis (4 June 2001), Maurice Kane (July 2001), Brown McDonald, Jr. (22 June 2001), Miller, Walker H. Powe, Jr. (16 August 2001), and Neel Price (June 2005).
46 P. H. Hostetter (1999) *Doctor and Soldier in the South Pacific* (Versailles, MO: B-W Graphics), "armloads" p. 14 and pp. 6–30.
47 Hostetter, *Doctor and Soldier*, pp. 14, 15; Questionnaire, "usual" McDonald, and Miller and Price.
48 Questionnaire, Price.
49 Hostetter, *Doctor and Soldier*, pp. 15.
50 Ibid., pp. 16.
51 FM 7–30, (1 June 1944) "Service Company and Medical Detachment (Supply and Evacuation) Infantry Regiment" (Washington, D. C.: Government Printing Office).
52 Hostetter, *Doctor and Soldier*, pp. 16
53 Questionnaire, McDonald. See Chapter 2 for additional detail on McDonald's training and initial combat experience.

2
Baptism of Fire

Abstract: *This chapter explicates the Army's shortsightedness, emphasizing combat initiation experiences of the both original and replacement first echelon medics. Each baptism of fire exposed the critical gaps in training programs, forcing the medic into a continuing on-the-job training mode. Even medical officers, participants in highly organized training programs, had to rethink their strategies and goals once they entered the combat environment. More critically, enlisted men who served as first echelon medical soldiers had not, as a group, enjoyed an appropriate training regimen designed to ready them for the physical and emotional rigors of caring for the combat wounded. Yet while the Army had profoundly under-prepared them to deal with the ghastly reality of combat, aid men adapted extraordinarily to treat the war wounded.*

Shilcutt, Tracy. *Infantry Combat Medics in Europe, 1944–45*, Basingstoke: Palgrave Macmillan, 2013. DOI: 10.1057/9781137347695.

Doc Wilbur Heinold, company aid man for E Company, 415th Infantry, marched for several hours with his untried unit across the Belgian frontier in late October 1944. As they moved toward the Dutch border, a sudden German artillery barrage caught them unprepared and the American troops dropped to the ground, trying to burrow in for the slightest protection. Although Heinold considered himself a "champion foxhole digger," his platoon sergeant tossed aside a foot of dirt and dived into the hole before Heinold could even put shovel to ground.[1] Shrapnel slashed all around them and despite the sergeant's efforts, a follow-up salvo struck and injured him.

Heinold leaped into the shallow pit to render aid and landed astride the wounded man who immediately seized his would-be savior by the throat. Heinold wrestled with the uncomprehending, injured soldier but could not free himself from the strong grip. Heinold's platoon leader rushed to help, prying the sergeant's fingers from the medic's neck. But by that time the sergeant lay dead, suffocated by a piece of shrapnel that had pierced his throat. At that moment Heinold understood that his training had been an illusion; combat would be its own teacher and "on-the-job training" had begun. Heinold's baptism of fire flung him headlong into a gory routine little related to his stateside training, a regimen that had consisted primarily of nursing sore feet.[2]

While not all front line medical personnel suffered so violent a combat initiation and while no training regimen could have anticipated each soldier's experience, first blood invariably reshaped each medic's understanding of his job and exposed the gulf between his preparation and the devastation of front line war. First echelon medical enlisted men who entered combat in Europe from the summer of 1944 to the spring of 1945 had participated in no standardized, uniform, or realistic program tailored specifically for grisly combat realities.

Some medical enlisted men claimed the benefit of limited medical training in hospitals or on maneuvers, but the initial enemy contact broke down their imagined world. Regimented, safe, and sanitary working conditions would not be factors in combat, nor would sore feet be their primary concern. Charged with salvaging those who could return to the line and evacuating those who could not, these medical soldiers generally benefited from the physical hardening of their training, but as a later analysis revealed, training for medical soldiers was deficient in opportunity to "make practical application" of any knowledge learned.[3] With first blood, medics comprehended that any medical skills that they possessed were woefully inadequate and must give way to new proficiencies gained in combat.

Other front line enlisted men entered battle with virtually no medical instruction, yet they too adjusted as best they could. While MC and MAC officers benefited from more consistent and thorough training, they also encountered surprises in their first combat experiences. For front line medics, the initiation to casualties revealed the superficiality of their training as they learned that caring for the combat wounded demanded continuing learn-as-you-go experience.

Medics who landed with their combat units in Normandy in the summer of 1944 carried no reflectively paced introduction to war, but rather rushed headlong into a myriad of violent encounters marked by staggering losses. Aboard the landing craft, seasick soldiers lurched against each other in the early morning hours of the invasion and searched the horizon for signs of the enemy. The sights and sounds of battle intensified as the vessels neared the beaches, reminding those on board that the "enemy fights back."[4] Straight ahead, red and orange explosions flashed in the gloom even as bullets pinged against the ships.[5] Dawn gave way to a misty morning and then to gray afternoon as enemy ordnance relentlessly pounded ships jockeying for position like "horses for races."[6]

The landing craft offered little protection to the soldiers on board, and casualties fell even before the men reached the shore. Enemy shells slammed into the crafts, and shrapnel forced the evacuation of some, while other vessels and their occupants simply vanished in fiery explosions. Onboard one transport, a combat unit watched in horror as the body of one of its own soldiers blew apart when a bullet detonated a hand grenade he had slung from his pack.[7]

Under this heavy fire the assault crafts maneuvered as close to shore as possible. Some stopped gently while others crashed into underwater concrete and steel obstacles or stuck fast on shallow sandbars.[8] In the prevailing confusion and panic soldiers poured from the landing crafts into the surf. Many plunged into water much deeper than anticipated and all struggled toward the shore as if in "slow motion," sinking under the weight of their gear. They fought to stay afloat in water that danced as enemy bullets sprayed across the surface. Cries of "I'm hit!" filled the air even as the wounded and the dead bobbed in the surf.[9]

Other voices joined in the din. Soldiers screamed as their equipment pulled them under and threatened to drown them. Officers shouted to thrashing troops to release the web-bound trappings. Men waded toward the beach leaving flame-throwers, bazookas, mortars, and ammunition to sink to the channel floor. As the soldiers reached the shore, those who retained their weapons returned fire, but casualties quickly mounted. Shell fragments tore through flesh while inland minefields slowed the advance of the "waterlogged" Americans.[10] Although the GIs moved across the French beaches as swiftly as possible, it seemed to them as if they walked into the "face of a real strong wind."[11]

The whirl of activity swept up the medical soldiers as they ministered to injured men still struggling in the water. On the shore, casualties mounted in what war correspondent Ernie Pyle called a "museum of carnage."[12] In the midst of the bloodletting, medics such as Private Dave Fought of the 1st Division labored endlessly without food or sleep. For three unrelieved days Fought moved doggedly from one casualty to another working under conditions "hard to describe."[13] He may have tended to the wounded more efficiently than most for Fought had trained "from the beginning" of his army service to work as a combat medic, even participating for almost six months in mock assault boat landings.[14]

Unlike most medics, this 18-year-old rookie drew confidence from his first aid instruction, directed by officers and enlisted men of the "Fighting First" who had been in the Africa and Sicily campaigns. Yet despite his faith in the "best possible" training, the mock landings and fake casualties had ill prepared him for the stresses that accompanied the combat job.[15]

Fought's initial duties in Normandy, as a member of a four-man litter bearer squad, seemed quite simple: move to a wounded man, place him on a litter, and haul him to a spot where he could receive life-saving medical attention. But Fought found that several factors complicated this maneuver, factors his trainers had not adequately emphasized: the dead weight of the wounded man, the harsh terrain, and the German soldiers who were doing their skilled utmost to kill them all. For Fought and other litter bearers, sand under foot compounded an already difficult task causing them to stumble. Caring for the casualty required that medics try to maintain a level stretcher, but they constantly lost footing in the sifting sand, crashing to the earth even as they tried to absorb the blows of the falls to keep their charges safe. Litter bearers soon gained the occupational scars common to their craft such as skinned forearms and shins, yet they continued their transports. On occasion bearers sprained or dislocated joints, or even snapped limbs and had to be reassigned or evacuated themselves.[16]

On the beaches, sniper and small arms fire peppered the ground, while charges blew, all hindering movement inland. The able-bodied soldiers left equipment and wounded behind, littering the sand and surf. The soldiers who managed to escape serious injury on shore trudged on to higher ground and reorganized once they located the scattered remnants of their platoons.[17] Fought's training and the fortunes of war kept him alive and allowed him to aid his fallen comrades, but even as he and his fellow medics continued caring for the wounded on shore, still more troops waded into the killing ground.

Aid man Frank Irgang shivered on a landing craft as he waited with his unit to disembark. His body shook from the icy sea spray leaping over the sides of the craft, the cold afternoon rain, and from the fear of the unknown. The men assigned to his care comprised an infantry heavy weapons company and although he had only recently joined them as a replacement aid man, camaraderie established on the brief trip across the Channel gave Irgang a feeling of solidarity in purpose. The man the unit called "Doc" had followed a bafflingly circuitous route before joining the 175th Infantry as a company aid man. And even though he had undergone stateside medical training, this knowledge failed to calm his fears as he anticipated the landing.[18]

Drafted in 1942, Irgang completed basic training at Camp Robinson, Arkansas. He then received instruction in general medical first aid, hospital ward duties, and records keeping, after which he transferred to the 88th General Hospital at Longview, Texas. While at the hospital, the Army accepted Irgang into the Army Specialized Training Program (ASTP), transferring him first to Texas A&M University to study psychology, and then to New Mexico A&M University for engineering training.[19]

After two university terms, the Army Air Corps assigned Irgang to Amarillo Army Air Base in Texas for yet another round of basic training, then sent him to the University of Nebraska to study math, physics, chemistry, geography, physiology, and navigation. He reported to Santa Ana Air Base in California but almost immediately he received new orders to proceed to navigator school in Hondo, Texas. Before he could transfer to Texas, however, the Army directed anyone with ground forces training to return to the infantry. Irgang boarded a troop train along with several hundred other soldiers and headed east.

At Camp Grant, Illinois he underwent a refresher course in medical care taught by "dedicated and knowledgeable pharmacists," and in the late spring of 1944 shipped out to England, assigned to a medical detachment.[20] After a brief stint with a BAS he transferred to the 175th Infantry and sailed for France with the "tough" men of D Company.[21]

When Private Irgang's boat crashed into an underwater barrier on D-Day, the officer in charge ordered the men into the cold, neck-deep water. Irgang jumped in and as he wrestled through the surf toward shore, he caught glimpses of waves rolling over fellow soldiers as they spilled into the water, desperately struggling to stay afloat under the weight of their heavy equipment. He spied a wounded buddy and grabbed the soldier, but as they labored landward Irgang plunged into a shell hole and lost his grasp, fighting the water. Panicked and gasping for breath, he found solid footing once more and waded alone for shore.[22]

Exhausted and still choking, Irgang located the remaining members of his unit, now reduced from 40 to 22. D Company worked its way inland through marked mine fields with Irgang bringing up the rear. The soldiers dug in for the night, and the cold, wet aid man moved "from hole to hole" assessing the welfare of his men before tending to a nearby platoon whose aid man had not made it.[23] Irgang's first contact with the enemy proved "totally different than the precision and organization of training," yet his role in this horrific initiation confirmed his fitness for the demanding duties. Irgang no longer harbored any illusions concerning combat; men in his unit would die, and his own existence was equally fragile.[24]

As Fought's and Irgang's assault units drove into the "slaughterhouse" of the hedgerows during the following days, behind them more infantry troops pushed directly into the chaotic beaches.[25] Awed by the debris of earlier losses, 2d Division troops passed barges laden with wounded soldiers headed for England. This sight sobered medic Allen Johnson who, in 1939, had joined the "Indianhead Division" having no idea about the "seriousness of a war or killing people."[26]

An orphan, the then 16-year-old quickly claimed the 9th Infantry as his family, finding a sense of home and belonging perhaps for the first time in his life. His initial "recruit training" at Fort Sam Houston, Texas bonded him even further with his fellow medical enlistees. They listened to dry lectures from doctors while the infantry drilled outside the classrooms. Johnson judged that overall, the training regimen was undisciplined because his cadre had medical doctors who served as their commanding officers. In addition, the infantry envied Johnson and his cadre because the medics could "get by with doing almost anything" or in some cases, doing nothing. On occasion, the medical soldiers accompanied the infantry on field exercises, but Johnson judged these "a pain in the butt" that had little practical application for enlisted medical personnel.[27]

Johnson's peacetime training experience ill prepared him for the devastation that he encountered as a 20-year-old aid man. As the 9th Infantry pushed through the hedgerows of France in June and July 1944, the drive proved costly; after 60 days of combat only one of the riflemen under Johnson's care had originally come ashore in Normandy with the unit. Having no biological family, the violent loss of his adopted brothers left Johnson remote and distant from the replacement soldiers who streamed through the 9th Infantry to war's end.[28]

Medics also joined the casualty lists during the first days of the invasion; like the fighting men, medics were wounded and killed and replacements filled the slots of company aid men. Private Charles Cross, one of the "mass of replacements" in Normandy on D-Day plus one, waited on the beachhead for assignment.[29] A sergeant barked the roll call of those destined for the 2d Division, and as Cross heard his name he followed several others up a hill to trucks bound for the nearby front lines. He joined the 2d Engineer Combat Battalion, and until he evacuated more than a month later, Cross's unit faced unrelieved combat. Cross found ready acceptance as a vital component of his unit, but he had virtually no connection with the medical unit that was supporting him and he grew increasingly frustrated at his lack of realistic preparation.[30]

As they moved through hedgerow country, Cross "made it [his] business" to cement relations with troops belonging to the 2d Battalion, but he found that the nature of his work actually isolated him. Cross cared for casualties quickly, and then scrambled to catch up with the advancing troops. He patched up the wounded, treated shock victims with injections of morphine, and moved on. At one point, a call rang out for a medic at the rear of a thick and overgrown nine-foot hedgerow. One section had been "blasted clean," and a .50 caliber machine-gun stood idle and unattended in the breach. Five soldiers lay sprawled on the ground, all "desperately wounded" and in deep shock. Cross approached them, taking in the injuries that he would have to tend. One man's arm hung "by shreds," and while all the GIs had multiple wounds, Cross thought there was surprisingly little bleeding. He bent over each soldier to determine whether any still breathed and though none of the men responded, Cross called for a litter team and moved out, certain that these wounded men would never make it to the aid station. This was not an unusual situation for the combat medic during his 34 continuous days on the line. Cross's dissatisfaction and aggravation with the evacuation system increased daily as his total disconnection from the litter bearers and the aid station left him not only isolated but also constantly doubtful that his first aid to the wounded served any purpose.[31]

One month into combat Cross received a leg injury but continued to shuffle along behind his unit. At one point the aid man paused, resting for a time just a few hundred yards behind the front. He was in an area enclosed by small hedgerows that blocked out the tumult of battle, and the paradoxical beauty and momentary silence transported Cross from the war. As he recovered in this temporary haven he looked up to see two German soldiers staring at him. Cross had his Red Cross brassards on and thought that symbol must have reassured them because they made no move toward him. One man had a wounded foot, and the other bled from both ears; neither showed a weapon. The American aid man cautiously approached them, bandaged the foot and cleaned the head injury, and then pantomimed that they should help each other back toward their own line. Relief overwhelmed Cross as he limped away from the Germans, but the incident reinforced his growing feelings of isolation. He rejoined his unit and continued giving aid until the day he walked into a German machine gun nest. Running for cover he re-injured his leg, gave himself a shot of morphine, and called for litter bearers. His service as an aid man ended with his own evacuation.[32]

The emotional stress of his ordeal defined Cross's brief combat experience, leaving him isolated even though his comrades accepted him as one of their own. The men of his unit often dug Cross's foxhole for him at night, an extraordinary homage hardened riflemen paid only to medics, and refused to move out without him. One platoon member who stayed with the unit through the end of the war judged Cross the "best medic I ever saw [and] I saw a lot of them."[33] Cross's service as a front line medic proved invaluable to the combat team, but the long days in the field with the wounded combined with his emotional isolation and the inadequacies of the backup system left Cross constantly frustrated.

As the second wave of Allied troops worked their way through the hedgerow country during the summer of 1944, they followed a widening trail of devastation. Many of the American dead had been removed, but German soldiers lay where they had fallen and dead farm animals decomposed in the fields, their bloated carcasses giving off a dreadful stench. A "constant stream of casualties" passed to the rear even as the fresh combat units raced to the front.[34] Replacement aid men, like Cross, continued to join blooded combat units, most of which had little or no break since the invasion.

Private T. William Bossidy caught up with the 1st BAS of the 9th Infantry soon after the breakthrough at St. Lô. Because he arrived late at night, station personnel told him to sleep in a hedgerow dugout; he would receive his assignment the next day. Bossidy squeezed into the dugout and spent an uncomfortable night pressed against another straw-covered occupant. When an aid station medic woke him up the next morning, a startled Bossidy found that a dead German had been his dugout partner. Other surprises awaited Bossidy who had trained stateside as a cook. He received immediate assignment as a litter bearer, and despite his lack of medical training, quickly took to the job, developing a close association with his fellow bearers as well as the battalion surgeon. He also spent time with a rifle platoon as a replacement aid man, earning the respect of his combat unit.[35]

While the D-Day invaders pressed deeper into the French countryside, other units continued their thrust into southern Europe, initiated by the earlier invasion of Sicily in July 1943 and then Italy in September 1943. By February and March, 1944, as the 88th Division slogged its way up the Italian boot, 351st Infantry's battalion surgeon William Powe felt that his "realistic and graphic" training at Carlisle Barracks had partially prepared him for combat, but the intensity of the wounds and the sheer numbers of casualties overwhelmed even this physician. In addition, Powe distressingly found that the time needed for administrative duties cut deeply into the quality of care for the troops. So, when replacement MAC officer 2d Lieutenant Ben Martinez arrived at the BAS in June 1944, he relieved the doctor of much of the organizational responsibilities, freeing Powe to "devote full time to treatment of casualties." Learning quickly, Martinez became Powe's "right arm," devoting his attentions to "running the BAS," and Powe credited the success of their combat partnership to this efficiency.[36]

As the battles surged across Europe and the summer months gave way to autumn, untried infantry troops and medical soldiers replenished the ranks of the veterans. Soldiers assigned to BAS followed the attack force closely, which often placed them directly in the path of fire, and the resultant mayhem sometimes worked with naiveté to disorient the replacement medics.[37] Indeed, unpreparedness often led to irksome initiations which the Army later judged as inefficient, suggesting that "it is imperative" to train more thoroughly replacement company aid men in tactics, as well as in the technical aspects of first aid. But, in the thick of war, such evaluation was secondary to the need to provide replacements and so the Army continued to assign under-prepared reinforcement medical soldiers to front line positions.[38]

Private Ben Burnett, drafted in 1942, numbered among the GIs reassigned from the ASTP to the 104th Division at Camp Carson in late March 1944. He joined the BAS, confident in his "medical background" that included hospital training. But Burnett lacked both the physical discipline imposed by basic training as well as a fundamental understanding of the function of the BAS. By the time the 104th Division sailed for Europe in late summer, he had benefit of only a brief introduction to BAS duties and the tasks of litter bearers.[39] Burnett remained fundamentally ignorant of his role as he entered combat, running up against a perplexing problem the first time he rendered aid to a wounded comrade.[40]

Burnett started with the BAS as a litter bearer as the unit raced through France, but he quickly moved to the line as a replacement aid man just as the unit closed with the Germans. Soon after Burnett joined his new company, two men from his platoon crossed into the German lines under cover of nightfall to silence a sniper who had "inflict[ed] severe damage" on the group.[41] The soldiers had been gone only a short time when one of the two ran back demanding a medic for his wounded buddy. Burnett realized with a start that he was the medic, but stood there "speechless," not knowing quite what was expected of him.[42] The aid man looked at his lieutenant for guidance since no one had told him that his duties might involve entering enemy territory. The officer merely gazed back at the medic.

After an uncomfortable silence, Burnett staggered forward because he "didn't know what else to do." He later commented that he did not act out of bravery, but responded because he knew his duty. The surviving patrol member directed Burnett toward his wounded comrade and as the rookie aid man walked upright into the night a bullet "whizzed" by his head. Burnett dropped flat to the ground and inched back toward the waiting platoon. As he wriggled across the open space, he "discovered the body" of the fallen scout. After a cursory exam in the dark, he found that the soldier was dead, then continued safely on his retreat.

Reflecting on the episode over a half century later, he judged that his training had prepared him to follow commands and to perform his job, but the lack of a directive from the lieutenant left Burnett isolated from a command structure for the duration of his service as a company aid man. Effectively on his own, he felt segregated from the combat unit as a whole and learned quickly that aid men walked alone.[43]

While entire units such as the 104th Division entered combat in the fall of 1944, individual replacements constantly joined veterans on the front, soldiers who had long since settled into a "trance-like numbness" that allowed them to continue forward regardless of consequences.[44] Replacement Paul Winson landed in France in October 1944 anticipating an assignment to a rear medical facility because a vision problem restricted him to limited service. But, "somewhere between Omaha Beach and the Siegfried Line" Winson signed a waiver releasing the Army from any responsibility should he lose his remaining good eye. He was on his way to the 30th Division as a company aid man.[45]

The day after Pearl Harbor, Winson had tried to enlist in the Navy but was rejected because of diminished vision in his right eye. But in August 1943, the Army drafted him, judging his sight adequate for medical duties. While in basic training at Camp Grant in Illinois, he was placed in a cadre with fellow New Yorkers who engaged in "intensive" infantry exercises without weapons and "some but certainly not adequate" first aid training, most of which ended in laughter. However, Winson appreciated that the first aid instructors at Camp Grant hammered on two items that proved invaluable in combat: induce breathing and stop bleeding.[46]

When Winson joined B Company of the 120th Infantry on the Dutch border, the unfamiliar sights and sounds of combat at first disoriented him; the aid man had "no idea" what all the noises meant, nor who all the soldiers were who milled around the area during a momentary lull. A veteran rifleman set him aright by pointing out B Company, dug in around some houses and identifying a group of men located about 100 yards away as Germans.[47]

Winson, clearly showing nervousness about his new assignment, found understanding in two new buddies in the 2d platoon. A sergeant who was "empathetic" to Winson's state of confusion suggested that a cup of coffee might help the new man relax, so Winson readily accepted the offer. The sergeant poured the coffee into a china cup procured from a nearby house, but before the veteran could hand Winson the cup, another "prankster" sergeant noted the medic's plight and suggested that the cup be placed on a saucer. Joining cup to saucer, the men passed the coffee to Winson. When the aid man finally took the drink in hand, the "noise from shaking could be heard in the German lines." The two battle-tried men "laughed like mad" and immediately bonded to their new medic and he to them. The resumption of the war soon doused the levity of the moment when Winson's platoon came under intense fire and the aid man answered his first call.[48]

The air around Winson screamed with incoming artillery and mortar fire as he made his way across a sugar beet field to help his first wounded comrade. The medic shoved aside his fear, knowing that he "had a job to do." He crawled so close to the ground that he "could have been mistaken for a mole" and the sugar beets' foliage, no more than ten inches high, seemed to him a screen effectively shielding him from any harm. Relying on the real or imagined protection of the plant growth, he reached the casualty. Winson continued hugging the Dutch soil while administering aid, willing himself invisible to the enemy. The medic survived this first combat encounter in large part due to skills he learned in infantry basic training. But there were "no autoclaves on the battlefield," and the medic developed rough and ready first aid skills as the 30th Division pushed across Europe.[49]

Winson's baptism of fire also laid the cornerstone for a psychological barrier that he erected to "maintain sanity."[50] The aid man simply believed that he would not be hit, and that conviction compelled him forward in the days to come. The "Boy Scout camp" atmosphere at Camp Grant and the professional environment of the Lawson experience ill prepared Winson for the carnage inherent to his job, forcing him to devise his own measures to cope with his surreal environment.[51] As Winson moved along with his combat unit, additional divisions joined the fight in Europe during the winter months, and rookie medical soldiers like Winson encountered not only the shock of the horrifically wounded, but also the increasing complications imposed by weather and terrain.

In November 1944, fresh troops joined the Vosges Mountains campaign, underway since mid-October. The steep, rocky environment of the 30-mile wide range made transportation difficult. Roads were poor —if they existed at all—which forced supply and evacuation convoys to rely on rough, make-shift trails; the wounded often had to be hand-carried over the hazardous routes while the enemy fired down on them from the heights. When one of the relief units, the 100th Division, went on-line in early November they not only had to deal with these terrain problems, but they also encountered extreme weather conditions.[52]

The Army's high command had miserably failed to equip the 100th Division to survive cold, wet weather and as the untested and shoddily equipped regiments of the 100th moved in to replace elements of the 45th Division, the sight of "tired, wet, bedraggled, and war weary" veterans shocked the green troops and exposed the environmental difficulties to soldiers, alerting them to their coming tasks.[53] With only combat boots and overcoats to fend off the elements, it would be some time before the soldiers received winter gear, condemning them to devastating non-combat injuries and diseases, primarily trench foot and cold-related illnesses.[54] While the linemen of the 100th Division dug in, a group of officers, "dumb to the ways of war," huddled together in a forest clearing. A more experienced battalion commander warned the clique to scatter, but before they could scramble for cover, an artillery volley dropped directly on their position. Their combat initiation was underway.[55]

Captain Brown McDonald, 3d Battalion surgeon of the 398th Infantry, numbered among the naïve officers fortunate enough to escape the blast. Pearl Harbor had found McDonald enrolled in medical school, having completed four years of medical Reserve Officer Training Corps training. Graduating in 1943, McDonald gained the silver bar of a 1st Lieutenant in the Army along with his M. D. degree. The physician completed an abbreviated nine-month internship and reported for active duty in the late fall of 1943. At Carlisle Barracks' Medical Field Service School he learned about the "function of tanks and infantry in the attack" during his three-month course of study. He also underwent a "gas chamber experience," learning the distinguishing odors, colors, and effects of various chemical agents.[56]

McDonald then spent two months at Walter Reed General Hospital preparing for service in the South Pacific by studying tropical diseases. He also worked on the Infectious Diseases ward with young soldiers who had contracted "childhood diseases" and with others who had viral illnesses, spending many "twenty-four hour plus" working days making rounds and examining patients. McDonald left the Washington, D.C. area in late spring 1944 for Fort Bragg, North Carolina to join the 100th Division, a unit destined for Europe's cold-induced problems rather than the tropical illnesses of the Pacific.[57]

The 100th Division, in training since late 1942, had most recently participated in maneuvers in the Cumberland Mountains. Within weeks of its completing these field exercises, the Army stripped the division of approximately 3,500 men, deploying these troops both to the Pacific and to Italy in an attempt to return seriously depleted fighting forces to strength. Throughout the late spring and summer of 1944 the skeleton staff of the 100th Division remained stateside and worked feverishly to prepare its thousands of newly assigned soldiers, including McDonald.[58]

At Fort Bragg, McDonald's aid station personnel joined the infantry troops on the infiltration course, crawling low as rounds of live fire zipped through the air just above their helmets. McDonald's medical soldiers rehearsed the evacuation process, following directives from the Division Surgeon who emphasized the need to "send the casualties to the rear installations as soon as possible."[59] The 3d BAS men also joined in "extensive class-room instruction" introducing them to the paperwork that followed the casualty through the system.[60]

Now awaiting his first casualties in the scarcely tropical Vosges Mountains, McDonald, in common with other battalion surgeons, placed false confidence in his medical training and skills as a healer. From his time at Fort Bragg, he did understand clearly the threefold mission BAS personnel directed toward casualties: "maintain the status quo, impede further deterioration," and quickly move them back through the system. The violence of the war zone, however, exposed the physician's inexperience both as a soldier and as a healer. A fast learner, McDonald at least avoided bunching up his men in open territory, and shortly after treating his first patients at the aid station, he sought out skilled aid men to help him perfect his bandaging abilities. His dressings tended to slip while those applied by the medics "stayed intact." So the senior medical man acknowledged that he had a "lot to learn" as he worked in concert with the members of his medical team. McDonald extended his training beyond the confines of pre-combat schooling, assiduously honing his soldiering and medical skills on the front during the months that followed his first encounter with enemy fire in a forest clearing at the foot of the Vosges Mountains.[61]

The 87th Division, like the 100th, entered the European theater in late 1944. Advancing to Metz, France, the troops moved "like sleepwalkers" across the winter countryside toward the division's objective, a German-held fortress which had been "under siege for some time."[62] An extended cease-fire provided the Americans the opportunity to relieve their weary countrymen, and rookie company aid man Richard Lease moved up with his platoon taking over an established outpost.

Once settled, Lease's platoon sergeant reconnoitered the area to check positions. But evidently he had not warned his own men, and one of them mistook him for a German and shot him. Lease responded and ably attended the sergeant; he had accompanied riflemen on the practice ranges during training and had treated similar injuries. While the medic probably anticipated caring for a buddy wounded by enemy fire, an accidental shooting served as Lease's "first taste of real war" and taught him that combat would present both surprises and challenges.[63]

In *Wartime*, Paul Fussell's unblinking analysis of World War II, the author reflects on his own experiences, using the role of the medic to emphasize the disparity between the hope that soldiers carried into battle and the brutal understanding born of their consequent combat experience. Fussell maintains that prior to their first battle experience infantrymen reasoned that combat would require minor attention from medical aid men, much the same as they had experienced in training. They anticipated that medics would handle foot and ankle problems and treat minor injuries. Fussell holds that "few could have gone on" had they fully understood the consequence of combat and the challenging role of the combat medic.[64]

Not only did infantrymen lack an understanding of war's brutality, but the medics' course of study and their training exercises also engendered equally unrealistic expectations within the medical troops. The sensory "mosaic" of the medic's first blooding—the confused movements, the crash of weaponry, the screams and labored breathing of the wounded, the mingled stench of cordite and death—had the power to immobilize.[65] Yet battle demanded immediate responses from men who had, at best, practiced on simulated casualties under controlled and unrealistic circumstances, if they had had any combat aid preparation at all. And the combat medics responded more courageously and effectively than any might have anticipated.

Because the front line medics participated in such varied training programs, and because each man's combat experience was individual and highly personal, it is implausible to suggest any single description as a typical mode. But certain clear trends emerge from medics' stories that advance a better understanding of the relationship between preparation and combat experiences.

The Army's haphazard approach to readying medical soldiers for combat duty emerged the first time a medic answered the call of a wounded comrade or tended an injured soldier in the BAS. While medical officers and some enlisted men did report satisfaction with their medical training, most often a combat medic's training regimen had taught him little more than proper military behavior and hardened him physically for the tasks ahead, while introducing him only superficially—if at all—to emergency first aid care and teaching him virtually nothing of how to plunge into the fires of the battle and apply his skills under a lethal hail of enemy ordnance.

Medics kept up physically with their troops because they had undergone the infantry training regimen of forced marches, bivouacs, and obstacle courses. As the war progressed, fieldwork during training took on increased importance, so that the hardening also included dealing with the environment. And even though training exercises never adequately replicated or suggested the severity of the challenges that any soldiers would face in combat, the Army absolutely failed to convey the harsh reality that medics would face on the combat line. The amphibious landing, simultaneous multiple casualties, and the rapid and bloody pace of the first days in Normandy pushed the medics' physical and emotional endurance beyond anything they had been trained to expect. The soldiers who faced winter in the Ardennes, the Vosges Mountains, Italy, and other hostile areas dealt with hazardous terrain and severe weather for which the Army had neither prepared nor equipped them. Replacement medical soldiers rarely had any idea of their impending task.

Beyond the job's daunting physical demands, however, emotional travail stalked the medic as he fought a private war even as he cared for the needs of those felled in the bloody conflict swirling around him. When he first came under fire, the front line medical soldier recognized his own mortality and knew immediately that in order to continue his weaponless soldiering he must forge some sort of psychological armor which might allow him to move effectively between the world of the living and that of the wounded and dying.

Private Earl Lovelace, company aid man with the 2d Division, captured the essence of the combat medic's grotesque experience as he told of his time in the field. As he went weeks without bathing, Lovelace's physical appearance progressively transformed into a ghastly absurdist work of art as body and clothing morphed into a blood-soaked canvas. The shades of crimson changed daily as new layers of gore added to the mélange, some of it his own, but most coming from the wounds of his companions. He "couldn't get away from it" and his training touched these elemental realities not at all.[66] In common with his fellow medics, naked fear remained a constant companion. As the medics waded through the carnage day after day, often only a numbed spirit allowed them to continue.

For such soldiers as BAS medics and litter squads, their close working relationships helped reduce the strain. But for the company aid men, the confused nature of their command structure, the job's inherent loneliness, and their singular psychological disengagement only heightened their anxiety, forcing upon them a profound and unique sense of isolation.

For all combat medics, the battle line forcefully initiated them into the reality of their role in the ongoing conflict. First echelon medical soldiers quickly recognized that the battleground's violence stood in stark contrast to previous expectations and training, and they understood that they must refine their skills in the field. Their initiation foreshadowed the incomprehensively demanding task that lay ahead and they accepted that combat meant wounds more violent than they had ever imagined, a physical environment they had not been prepared to deal with, and emotional turmoil that would not slacken. Yet despite their terribly flawed training, combat medics learned rapidly, adapted and persevered as they overcame the first shock of battle, fully aware that their practical education had just begun.

Notes

1. "champion" Questionnaire, Heinold; L. A. Hoegh and H. J. Doyle (1946) *Timberwolf Tracks: The History of the 104th Infantry Division 1942-1945* (Washington, DC: Infantry Journal Press), pp. 42–66.
2. Questionnaire, Heinold; Hoegh and Doyle, *Timberwolf*, pp. 42–66.
3. *Report of the General Board* "Medical Section Study Number 88," p. 7, MHI.
4. F. J. Irgang (1949) *Etched in Purple* (Caldwell, OH: The Caxton Printers Ltd), p. 3.
5. Interview, "Private Elmer E. Matekintis Company F, 16th Regiment, 1st Division, D-Day, H Hour," Hospital Interviews (HI), Box 24242, RG 407, NARA.
6. Interview, "Captain C. N. Hall, Assistant Battalion Surgeon, 2nd Bn., 16th Infantry, 1st Division," HI, Box 24242, RG 407, NARA.
7. Interview, "Hall" RG 407, NARA; Irgang, *Etched*, p. 3.
8. Interviews, "Hall" and "Matekintis" RG 407, NARA; Irgang, *Etched*, p. 3.
9. Interview, "Matekintis" RG 407, NARA.

10 Interview, "Lieutenant John Spaulding, Leader 1st Section, Company E, 16th Infantry, 1st Division, D-Day Landing, by Master Sergeant. F. C. Pogue and Staff Sergeant J. M. Topete, 19 February 1945," EC.
11 Interview, "Spaulding by Pogue and Topete," EC.
12 Irgang, *Etched*, p. 4; "museum" E. Pyle (1944) *Brave Men* (New York: Henry Holt), p. 366.
13 Questionnaire, David E. Fought (7 April 2001).
14 Questionnaire, Fought; R. W. Baumer and M. J. Reardon (2004) *American Iliad: The 18th Infantry Regiment in World War II* (Bedford, PA: The Aberjona Press); United States, Army, 1st Division (1995) *First Infantry Division World War II: The Big Red One* (Paducah, KY: Turner Publishing Company).
15 Baumer and Reardon, *American Iliad*; United States, *First Infantry Division World War II*; D. Fought, "Memories of a 1st Division Medic in World War II," *Bridgehead Sentinel* Society of the First Infantry Division (Summer 2000), p. 6.
16 Questionnaire, Fought.
17 Interviews, "Matekintis" and "Lieutenant Clarence Akley, Company M, 9th Infantry Regiment, 2nd Infantry Division," HI, Box 24242, RG 407, NARA.
18 Irgang, *Etched*, pp. 2, 3; Questionnaire, Frank J. Irgang (March 2001).
19 Irgang, *Etched*, pp. 2, 3; Questionnaire, Irgang.
20 Ibid.
21 Irgang, *Etched*, pp. 1–3; Questionnaire, Irgang.
22 Irgang, *Etched*, p. 4.
23 Ibid., pp. 4, 5.
24 Questionnaire, Irgang.
25 R. Casey (1945) *This Is Where I Came In* (New York: Bobbs Merrill), p. 173.
26 Questionnaire, Allen L. Johnson (n.d.); United States, Army, 2d Division (1979, 1946) *Combat History of the Second Infantry Division in World War II* (Nashville, TN: Battery Press).
27 Questionnaire, Johnson.
28 Ibid.; See Chapter 3 for details on Johnson's experiences in the hedgerows.
29 Questionnaire, Charles C. Cross (10 June 2001).
30 Questionnaire, Cross.
31 Ibid.
32 Ibid.
33 Ibid.; "best" Sergeant Louis J. Bolla to Charles C. Cross, 30 June 1945, in Cross's possession.
34 R. B. Bradley (1970) *Aid Man!* (New York: Robert Bradley), p. 50.
35 Questionnaire, T. William Bossidy (1 July 2001). See Chapter 3 for details on Bossidy's baptism of fire.
36 All quotes Questionnaire, Walker Powe (6 August 2001); J. P. Delaney (1946) *The Blue Devils in Italy: A History of the 88th Infantry Division in World War II* (Nashville: The Battery Press).

37 Questionnaire, Frank Miller (2 April 2001); Hoegh and Doyle, *Timberwolf*.
38 *Report of the General Board*, "Study Number 92," MHI.
39 Questionnaire, Ben Burnett (3 July 2001); Ben Burnett to author, 8 March 2001, in author's possession; P. Mansoor (1999) *The GI Offensive in Europe: The Triumph of American Infantry Divisions, 1941–1945* (Lawrence: University Press of Kansas), p. 79.
40 Questionnaire, Burnett; Burnett to author, 8 March 2001; Burnett to author, 24 March 2001, in author's possession; Burnett to author, 27 April 2001, in author's possession.
41 Burnett to author, 27 April 2001; Questionnaire, "inflicted" Burnett.
42 Questionnaire, Burnett.
43 Questionnaire, Burnett; Burnett to author, 27 April 2001; Burnett to author 28 April 2001, in author's possession.
44 Questionnaire, Walter Biggins (29 June 2001).
45 Questionnaire, Paul Winson (31 May 2001).
46 Questionnaire, Winson; See Chapter 1 for details on Winson's training. Post-war analyses revealed that medical replacements were "never available in sufficient quantity." Poorly trained medics, including Winson, were the answer to this critical shortage—a problem they should have foreseen from World War I. *Report of the General Board*, "Study Number 88," p. 5, MHI.
47 Questionnaire, Winson.
48 Ibid.
49 Ibid.
50 Ibid.
51 Winson to author, 11 July 2001, in author's possession.
52 Questionnaires, Carl R. Aschoff (14 June 2001) and Victor Nash (21 May 2001); K. E. Bonn (1994) *When the Odds Were Even: The Vosges Mountains Campaign, October 1944–January 1945* (Navato, California: Presidio Press), pp. 13–28, 70, 108.
53 Questionnaire, Aschoff.
54 Ibid. See Chapter 5 concerning the problem of trench foot.
55 Questionnaire, Aschoff; Questionnaire, "dumb" Brown McDonald (22 June 2001); L. Atwell (1958) *Private* (New York: Simon and Schuster), p. 42; S. Stouffer et al. (1949) *Studies in Psychology in World War II, Volume II* (Princeton: Princeton University Press), p. 278.
56 Questionnaire, McDonald.
57 Ibid.
58 Aegis Consulting Group, ed., "The 100th Division Association," available from http://www.100thww2.org/; Internet, accessed 30 July 2002; Questionnaire, Angelo Zanin (n.d.).
59 Questionnaire, McDonald.
60 Ibid.

61 Ibid.
62 Atwell, *Private*, p. 15; Questionnaires, "under" Richard L. Lease (18 May 2001) and Thomas Hoke (18 June 2001).
63 Questionnaire, Lease.
64 P. Fussell (1989) *Wartime: Understanding and Behavior in the Second World War* (New York: Oxford University Press), p. 19.
65 "mosaic," B. Phipps (1987) *The Other Side of Time: A Combat Surgeon in World War II* (Boston: Little, Brown and Company), p. 78.
66 Questionnaire, Earl Lovelace (31 August 2001).

3
Combat Reality

Abstract: *This chapter focuses on the company aid men and their function during combat. Operating as the first medical contact for the wounded men of their companies, and often for those of other companies within their reach, aid men continued their uniquely solitary on-the-job training. They encountered unexpected carnage and worked amidst un-exampled violence that compounded an absurdly difficult job; yet they adjusted to combat conditions by abandoning or radically modifying prescribed medical care doctrine.*

Shilcutt, Tracy. *Infantry Combat Medics in Europe, 1944–45*, Basingstoke: Palgrave Macmillan, 2013. DOI: 10.1057/9781137347695.

Doc, as his unit called him, pressed into the earth as he crawled toward a wounded rifleman. German bullets slashed overhead in the darkening battleground and when he finally reached his buddy he knew he would have to work blindly as snipers fired at any light. His initial assessment indicated that he was dealing with a chest wound. So without benefit of sight, Private T. William Bossidy thrust grimy hands inside his comrade's shirt, following the sticky trail of welling blood to the torn cavity. He could do no more than sprinkle sulfa powder in the wound and staunch the flow of blood before moving on to help other casualties.

Doc Bossidy had in fact no education in the medical arts; his journey to the front as a replacement aid man for the 2d Division reflects the often disjointed yet pragmatic ways in which the Army cared for its combat wounded. His schooling in the mysteries of Spam and eggs at Camp Pickett provided absurd preparation for the lifesaving duties facing him as a company aid man.[1]

As the struggle for Europe intensified, the flood of casualties demanded additional front line medical personnel. In response, the Army pulled Bossidy and others like him from kitchens and typewriters, tasking them first with litter bearer's work. But as the American forces pushed toward the Siegfried Line attrition among company aid men forced the erstwhile cooks and clerks to move beyond transporting the wounded to treating their injured and dying comrades. Following cursory field explanations of bandaging and morphine injections, Bossidy donned the brassards of a front line company aid man. As an initial caregiver, his new duties seemed deceptively simplistic: control bleeding, minimize shock, manage pain. But as he tended to his fellow soldiers he realized how external forces profoundly complicated his seemingly straightforward job.[2]

Bossidy's story, while distinctive in its details, reflects in its whole the chaotic world of each company aid man as he confronted his own unique circumstances. As the first medical contact for the wounded, aid men worked independently of the BAS medical soldiers, traveling with and living among the infantry platoons. Whether trained in the United States or, as in Bossidy's case, simply initiated on the field, company aid men in European campaigns learned that Army doctrine and training scarcely prepared them for the horrific realities of war. Successful medics rapidly adapted to combat conditions, insuring their own survival as well as that of their comrades. Learning on the go and under fire, they abandoned or radically modified prescribed medical techniques, discovered ways to utilize the changing terrains for their own protection and for the safety

DOI: 10.1057/9781137347695

of the wounded, and coped with unanticipated long-term problems. Endowed with a spirit of pragmatism, Doc Bossidy, and the thousands of "Docs" like him, performed their crucial role in the crusade to liberate Europe despite the inadequate, or even absent, training.

Stateside Army instructors had instilled into medical trainees the belief that medics could always maintain a well-stocked medical kit, replenished through regular visits to the BAS. Campaign realities made restocking difficult at best, requiring that aid men improvise when allotments ran short. Forced by necessity to stay with his platoon, the company aid man rarely knew when he might restock his kit, what devastation he might encounter, or how he might have to adapt. The rapid and erratic pace of battle and the demands of treating multiple casualties under frenzied conditions could quickly exhaust the medic's finite supply of bandages, splints, morphine syrettes, and sulfa powder, and many of the taught procedures failed under war's harsh imperatives. Typical of such impracticalities, the bureaucratic emergency medical tags (EMT) became combat's first victim.[3]

Theory dictated that the field medic fill out a tag, indicating the casualty's name and unit, and initial care given. The aid man must then attach the EMT to the wounded soldier as an essential record of treatment, informing the medical personnel behind the line. But company aid men promptly discovered the folly of taking time to fill out little cards in the midst of flying bullets. Some medics at first improvised by scrawling abbreviations of treatment on the EMTs, but by and large company aid men simply abandoned the formal process. Yet aid men understood the critical need to provide information to those who followed, so many devised crude codes: an "M" traced in blood on the casualty's forehead or a morphine syrette tied to his clothing signified that the soldier had received morphine, while a "T" shaped strip of tape on the forehead might bear marked numbers indicating the time of a tourniquet's application.[4]

At the same time, a rapid line advance demanded that medics limit their load to the most critical medical stores, so as the first troops struck into the French countryside many company aid men abandoned field packs and gas masks along the combat trail while some judged even their mess kits unnecessary. One 2d Division company aid man's approach to mealtime reveals his resigned acceptance of the horror that suffused his life. When he could take time to eat, this medical soldier simply wolfed down his daily rations straight out of his cupped hands, noting that this served a dual purpose: he could eat quickly, and at the same time, clean the dirt and gore from his hands.[5]

The speed of the offensive across France also meant that a aid man maintained at best a tenuous connection to his BAS. On those occasions when aid men accompanied seriously wounded back to support areas, they filled their medical pouches with whatever could be promptly acquired and then rushed back to the line. Some company aid men ordered runners to retrieve supplies and still others urged litter squads to bring up needful items.[6]

Desperate to husband their own precious cache, frontline medics used the wounded men's personal-issue aid kits when possible. These kits included sulfa powder and bandages. During engagements with heavy losses, aid men spent little time with ambulatory casualties, pausing only to direct the wounded soldiers to tend to their wounds and evacuate themselves. As the aid men scrambled to keep pace with their fighting units, they often instructed the walking wounded to report the locations of the more seriously injured to the litter bearers.[7]

Although basic military medical training hammered fundamental techniques for controlling bleeding, managing pain, and minimizing shock, the implacable reality of combat forced aid men to improvise lifesaving techniques and nowhere is this better illustrated than in medics' approach to devastating chest wounds. A direct bullet strike sometimes caused chest trauma, but more often the shrapnel from exploding shells tore into the chest cavity, compromising the wounded man's breathing. The wounded man might inhale, but air entering the chest cavity through the wound would not allow the lungs to expand. For such chest wounds, the aid man first had to seal the damaged area. This, according to stateside instruction could be done by covering the sucking wound first with a regulation bandage and then with a patch cut or torn from a raincoat.[8]

This technique could work, but the aid man might not have a raincoat, or the advised protocol might fail to decompress the lung. In such cases, medics experimented with innovative remedies. One aid man used an intact raincoat as a large binding around the wounded soldier's entire body to create an effective seal; another used mechanic's engine tape, finding that it tightly adhered bandages to the skin. One replacement aid man, bereft of any medical training, simply laid the handiest heavy object, such as a rock, over the wound to seal the damaged lung area. Not prescribed in any medical manual, and often not in keeping with the sterile protocol, such rough and ready treatments did save soldiers' lives.[9]

While in some cases combat medics had to invent new procedures, in other instances they discovered that army-issued medical equipment was woefully inadequate. In Italy, Private Russell Redfern of the 88th Division found that the free-bleeding arterial wounds presented him with the greatest challenge of his job. He learned that government issue (GI)[10] uniform belts made far better tourniquets than those in his medical kit. In addition, on more than one desperate occasion Doc Redfern simply thrust his fingers inside wounds to restrict the flow from the arteries until a shoelace or other handy substitute could be located to tie off the hemorrhaging vessels.[11] In common with Redfern, replacement medic Private John Sullivan found standard tourniquets useless. These would not fit around hips and buttocks so Doc Sullivan regularly supplemented his issued equipment with plundered German supplies to save American lives.[12]

Medics also developed methods for simultaneously coping with multiple casualties. In the French hedgerow country, Doc Bradley of the 30th Division prioritized with a mere glance at the wounded soldiers' skin and fingernails. If he noticed the "grey-green color of death," Bradley limited his consideration to gentle words of comfort and turned his healing attention to the casualties that he thought might have the best chances to live.[13]

Combat rarely allowed careful, thoughtful treatment, and in common with Bradley, aid men not only improved their field craft skills to aid their comrades more effectively, but they also learned the hazards and havens provided by ever-changing local terrains. The fierce attacks and counterthrusts in the French hedgerow country proved deadly to the forward units ordered to "keep moving whatever the cost."[14] German snipers systematically picked off American soldiers, while heavy artillery and mortar fire indiscriminately killed or maimed others.[15] Aid men remained constantly busy, moving from one injured man to the next, and within any given encounter they might treat any combination of wounds including chest, extremities, and face, as well as traumatic amputations.[16] Nighttime rarely brought relief as "artillery and mortar fire rained" on the troops, and patrols resulted in still more substantial losses.[17]

One problem marking the hedgerow country was the aid man's ability to get to casualties. Knotted, entangled growth bordered each parcel of land, creating formidable walls rising several feet above their roots in earthen dikes. These living green barricades separated German troops from American soldiers but offered only limited protection, and the enemy's close proximity meant that the aid men constantly worked under fire while caring for the wounded. So the medical men limited their exposure by holing up in protected areas until a cry for help rang out.[18]

2d Division aid man Allen Johnson used "angles and tricks" both to stay alive and to help his wounded brothers during the ordeal of the hedgerows.[19] As the American soldiers penetrated deeper into France, they found gaps in the hedgerows that allowed access to individual fields, but German soldiers took tactical positions behind the hedgerows, firing at GIs moving past these openings.[20] Johnson watched as the Germans cut down his rifle platoon members one by one. Then he noticed that the men fell to enemy fire in a fixed pattern; the Germans were waiting to shoot about every third or fourth man. The medic then listened for a shot to ring then sprinted across the gap.

Another time Johnson watched with disbelief as a fellow medic crashed to the earth, hit in the head by enemy fire. The wounded medic lay exposed on the field, calling to Johnson for help. Johnson pulled out a "calf rope" which he had picked up at some point in the push through the French countryside, and under the continuing fire, he dived out into the field, looped the rope around the injured man's foot, and scurried back to the safety of the embankment. Once there he pulled the wounded medic over to him but despaired when he saw the injury. The Germans had caught his fellow medic "right between the eyes." The injured medic died the next morning despite Johnson's efforts.[21]

The inventive Doc Johnson also devised a noteworthy technique for reaching wounded men who lay on the German side of the hedgerows. He agilely climbed the hedgerows and skimmed the tops of the green barriers "like a snake" with his body pressed as flat as possible in an effort to avoid making himself a target. The 30th Division's Doc Bradley developed another unique maneuver by perfecting his skill of diving through the hedgerows, landing in a somersault, flattening out, and then scrambling from incoming fire.[22] Apart from medical skills, company aid men clearly required a degree of imaginative acrobatic talent.

Field medics, as they adapted to the physical environment of the hedgerow country, aided well their fallen comrades but any consistency they developed in tending to the wounded here vanished in the new pandemonium of forest combat. From September to mid-December

1944, the Army committed nine infantry and armored divisions into the Hürtgen Forest. GIs dubbed this 70 square mile area of wooded terrain the "Green Hell".[23] The region nurtured foliage so dense that it obstructed the sun—what little time it shone—leaving the troops to fight in a gloomy shadow-world. Restricted visibility, limited at times to a few yards, combined with lethal mines carpeting the forest floor to further frustrate and compound the aid man's job.[24] Mine detectors proved worthless in locating these wood encased mines, and because each device contained just enough TNT to "blow off a man's foot or leg," medics attended a disturbing number of mutilating foot and leg wounds.[25]

Even as infantrymen tried to avoid the explosions from the ground, airborne missiles struck from above, as trees burst like jagged fireworks. German mortars blasting into the thick tree canopy splintered branches and limbs into a lethal rain of wooden projectiles, hurling death on the American troops below.[26] Once again, the medics' training had not equipped them for treating such wounds. Aid men moved with their platoons through shattered trees and dug in with their combat units, all working under horrendous conditions. One aid man's response to the complex nature of rendering aid in the Hürtgen reflects the medic's unspeakable task; he referred to the forest as a "death factory," but rather than elaborate on details, he understated his duties as "extremely difficult."[27] Another soldier's observations validate the medical man's hesitancy to elaborate:

> You can't get all of the dead because you can't find them, and they stay there to remind the guys advancing to what might hit them. You can't get protection. You can't see. You can't get fields of fire. The trees are slashed like a scythe by artillery. Everything is tangled. You can scarcely walk.[28]

Stateside training posited combat medics involved only in short-term medical actions. Advancing from field to field, foxhole to foxhole, it was imagined that they would administer first aid and immediately send the critically injured back to the BAS for more focused care. The shortsighted planners did not anticipate that the Army might become mired in a static winter's campaign of rain, mud, sleet, snow, and bone-shattering cold. War's demands meant the poorly trained aid men must assume responsibilities for additional medical problems, and once again the medics adapted, providing some solace to foot soldiers pressed beyond human limit in an otherwise desperate situation.

As 1944's fall chilled to winter, temperatures in the Hürtgen settled to the freezing mark, and intermittent rain, snow, and sleet turned the ground into a quagmire. Aid men lived with their rifle companies in foxholes that filled perpetually with water, ice, or snow.[29] The damp, frigid conditions multiplied the numbers of casualties due to trench (immersion) foot. At the same time, the emotional stress of unrelieved battle in the dark, hellish world of the forest added combat exhaustion to the growing list of debilitating non-combat elements. As the point medical soldier assigned to his combat unit, the aid man's burden expanded from tending combat wounded to include primary care-giving responsibilities for cold- and stress-induced injuries.[30]

Weather-related problems were of course not limited to the Hürtgen, as climactic conditions impacted every battlefield across the European continent. During late 1944 and early 1945, troops fought in a once pristine European landscape that combat had transformed into a wintry nightmare where the blood of the wounded dropping on "newly fallen snow" marked the path for those behind.[31] Linemen blasted through frozen ground with TNT to dig their foxholes, but combat soldiers found no relief from the cold as they burrowed into their icy trenches and gathered their blankets around them, hoping for some measure of warmth.[32] The army's abysmally neglectful failure to outfit the front line soldiers adequately with winter gear meant that riflemen contended with constant illness along with frozen hands and feet.

From December 1944 to January 1945, the numbing cold and unrelieved fighting in the Battle of the Bulge led to intense suffering among the front line troops and medics stayed constantly busy caring for the sick, as well as tending to combat casualties.[33] One medic admitted that he wondered often about the feasibility of purposely exposing his hand to frostbite as a means of escape.[34] Such pressures put the company aid men in untenable positions, and in the Ardennes litter bearers looked on the job of company aid man as a "death sentence."[35] They knew that as litter bearers they had to make dangerous hauls in the snow, but the relatively warm protection of the BAS seemed an exceedingly better alternative to the tortured existence of the foxhole medics. In addition to winter weather, which complicated the medic's already daunting task, night fighting further challenged aid men.[36]

When Private Francis Berry joined the 88th Division in Italy as a replacement company aid man, his battalion surgeon warned him that he would likely tend to casualties during night fighting. Berry could not fathom how to treat the wounded in blackout conditions and asked the doctor to elaborate. The officer assured the neophyte that there would "be plenty of light" for his purposes, but Berry found that he never had enough light to work properly. On one occasion Doc Berry's unit waited until after sundown and divided into two advancing columns. As he brought up the rear, Berry heard a burst of gunfire and within seconds one of his platoon members ran back, calling for help. Whispers to "keep down" drifted from the line as Berry approached the point, so the medic crawled into a field, finally locating the fallen soldier who repeatedly told the aid man that he could not breathe. Before Berry could strip the man of his gear, the soldier died. An investigation later determined that there had been no enemy close by and "he had been shot by one of [the unit's] own men."[37]

Other medics, like Doc Earl Lovelace, found success in treating combat wounded in the dark. During night fighting in France, Lovelace watched the shadow figures of a two-man patrol move out beyond the confines of a hedgerow to check on the German position. Within a short time shots rang out and one of the men soared back over the top of the hedgerow. The other soldier lay wounded, passionately wailing for his mother and unable to return. Despite the night's inky blackness, the aid man scrambled over the hedgerow, located the wounded soldier by moving toward the noise, and dragged the casualty back to a spot where platoon members brought him to safety. The medic tented the casualty with raincoats, and with the light blocked he treated the wound by flashlight.[38]

Whether working in daylight or in darkness, the company aid man carried no special immunity from the same weapons that felled his comrades. In the extreme, divisions with as much as six months of "severe combat" replaced over 100 percent of their aid men.[39] Medics, like their foxhole counterparts, "never knew when their time was coming."[40] In one unusual instance, a platoon lost its aid man but failed to receive a replacement medic, so Joseph Cosby, who carried the unit's Browning Automatic Rifle (BAR), took up the aid man's kits and helped his wounded companions even while fulfilling his original combat obligations.[41]

As company aid men moved along the combat trail, they quickly realized they had no control over an environment with carnage as the most elemental truth. But their creative adjustments allowed them to meet their fundamental duty of treating their comrades. When their issued kit lacked effective provisions they utilized nearby objects, snatched up German stocks, or used their own hands to treat injuries that they could never have imagined. The immediacy of saving lives overrode all previously learned procedures, and because they did their job even under the most singular of circumstances, most of the men they treated moved back through the evacuation system to receive more definitive care. For those mortally wounded soldiers, the company aid men could offer only gentle words of comfort.[42]

Front line medics, woefully trained, treated wounds they had not anticipated under extraordinarily challenging conditions, and although the landscape of the combat zones changed as the war progressed, the deadly chaos never slackened. For the infantry soldiers, company aid men represented the hope of survival that would link the wounded to the fragile life chain. Whether the combat medic forged this link in a rapidly advancing front or a static entrenchment, the medic adapted and improvised to fulfill his unspoken vow to his platoon, "until death ... do us part."[43]

Notes

1. Questionnaire, T. William Bossidy (1 July 2001). Having trained at one of the MRTCs it is possible that the Army assigned Bossidy as a medic simply by virtue of an assumption that soldiers trained at the MRTCs had some basic medical knowledge. However, Bossidy reports he received instruction as a cook following basic training at Pickett. He had no medical training until he was sent to a rifle company as an aid man. Bossidy indicated that even though he trained at an MRTC, his training had "no relationship" to the duties of the combat medics. UAR, "Camp Pickett, 1943," RG 112, NARA.
2. Questionnaire, Bossidy. Stateside training records indicate that the increased pressure to train greater numbers of troops in shorter periods of time resulted in assignments within the Medical Department that, like Bossidy's, appeared "crazy." Records further indicate that inexperienced officers who had no appreciation for the role of medics were commanding medical training regiments. In addition, there were reports of electricians training as mechanics, mechanics training as electricians, etc. Medical soldiers, in general, were given "insufficient explanation to make clear ... the need for and objectives of the work at hand." "crazy," UAR, "Camp Ellis, Illinois 1944," RG 112, NARA and "insufficient," "Psychiatric Consultation Division, Medical Training Section Fort Lewis Washington 22 August 1944," Box 222, RG 112, NARA.
3. FM 21–11 (1943) "Basic Field Manual: First Aid for Soldiers" (Washington, DC: Government Printing Office), p. 2; R. B. Bradley (1970) *Aid Man!* (New

York: Robert Bradley), p. 63; Questionnaires, Gerald W. Allen (25 March 2001), Roy Barratt (13 June 2001), William Braunhardt (30 May 2001), Frank R. Ellis (4 June 2001), Wilbur Heinold (2 April 2001), and Allen L. Johnson (n.d.); L. Atwell (1958) *Private* (New York: Simon and Schuster), p. 172. Aid kits usually had a collection of compress and gauze bandages, tape or adhesive plaster, splints, scissors, morphine syrettes, iodine swabs, sulfur powder, and aspirin.

4   Questionnaires, "M" G. Allen, and Ellis, Heinold, Frank Miller (2 April 2001), and Donald Warner (16 April 2001); Bradley *Aid Man!*, p. 49; "Unit History, 63d Infantry Division, Office of the Surgeon, 30 June 1945," Box 11162, RG 94, NARA; G. A. Cosmas and A. E. Cowdrey (1992) *The Medical Department: Medical Service in the European Theater of Operations* (Washington, DC: Center of Military History), p. 363; "T" F. J. Irgang (1949) *Etched in Purple* (Caldwell, OH: The Caxton Printers, Ltd.), p. 7; C. Whiting (1989) *The Battle of Hürtgen Forest* (New York: Orion Books), p. 74; TM 8-220 (5 March 1941)"War Department Technical Manual: Medical Department Soldier's Handbook" (Washington, DC: Government Printing Office), p. 143.

5   Questionnaires, "wolf down" Johnson and Everett Smith (7 August 2001); "Unit History, 63d Division," RG 94, NARA; W. S. Tsuchida (1947) *Wear It Proudly: Letters* (Berkeley: University of California Press), p. 38; Bradley *Aid Man!*, pp. 51–62; R. L. Sanner (1995) *Combat Medic Memoirs: Personal World War II Writings and Pictures* (Clemson, SC: Rennas Productions), p. 100.

6   Questionnaires, G. Allen, Bossidy, Braunhardt, Heinold, Johnson, Earl Lovelace (31 August 2001), and Russell Wade Redfern, (1 June 2001).

7   Questionnaires, Walter Biggins (29 June 2001) and Heinold; FM 21-11 (1943), pp. 87–111; Bradley, *Aid Man!*, p. 51.

8   Questionnaires, Francis Irvin Berry (29 May 2001) and Richard L. Lease (18 May 2001); Bradley, *Aid Man!*, 63; FM 21-11, p. 13.

9   Questionnaires, Berry, Johnson, Lease, and Lovelace; Bradley *Aid Man!*, 63; FM 21-11, p. 13; TM 8-220, p. 128.

10 GI is a nickname for American Soldiers.

11 Questionnaire, Redfern; FM 21-11, 10; TM 8-220, pp. 128–129.

12 Questionnaire, John T. Sullivan (n.d.).

13 Bradley, *Aid Man!*, p. 59.

14 Interviews, "keep moving," "Lieutenant Clarence Akley, Company M, 9th Infantry Regiment, 2nd Infantry Division," HI, Box 24242, RG 407, NARA and "Private Elmer E. Matekintis Company F, 16th Regiment, 1st Division, D-Day, H Hour," HI, Box 24242, RG 407, NARA; Interview, "Sergeant Willie L. Murray, Company L, 9th Infantry, 2nd Division," HI, Box 24242, RG 407, NARA; Interview, "Captain Edward W. Martin, Company C, 22d Infantry, 4th Division, by Lieutenant Stockton, 217th General Hospital," HI, Box 24240, RG 407, NARA; Interview, "Lieutenant Benjamin W. Mills,

Commanding Officer, and 1st Sergeant Lawrence E. Houck, Company F, 8th Infantry, 4th Division, 18 August 1944," Box 24021, RG 407, NARA; M. Doubler (1994) *Closing With the Enemy: How GIs Fought the War in Europe* (Lawrence: University Press of Kansas), pp. 41–60.

15 Interviews, "Akley," "Murray," and "Matekintis," RG 407, NARA; Interview, "Lieutenant William Anderson, Battalion S-2 and Lieutenant James Piper, Assistant S-3, 2d Battalion, 12th Infantry Regiment, by Lieutenant Fife, 18 August 1944," Box 24021, RG 407, NARA; Interview, "Lieutenant Colonel Linwood McClure, 112th Infantry, 28th Division by Lieutenant Posvar, 186th General Hospital," HI, Box 2421, RG 407, NARA; Interviews, "Action of Company E, 22d Infantry, 4th Infantry Division, 25 July to 2 August 1944, at Rocherath Germany, 24 October 1944 with 1st Lieutenant George D. Wilson, 2d Platoon Leader, Technical Sergeant John A. Cznadel, 1st Platoon Sergeant, Staff Sergeant Glen L. Phearson, 2d Platoon Sergeant, Technical Sergeant Michael Elenchik, 4th Platoon Sergeant," Folder 31, Box 24021, RG 407, NARA.

16 Interview, "Murray," RG 407, NARA.

17 Interviews, "artillery and mortar fire rained" "Anderson and Piper by Fife," and "Akley," RG 407, NARA.

18 Questionnaires, Braunhardt, Heinold, and Johnson; G. S. Johns (1958), *The Clay Pigeons of St. Lô* (Harrisburg, PA: Military Service Publishing Company), p. 3.

19 Questionnaire, Johnson.

20 Ibid.; Interview, "2d Lieutenant Theodore Ray, 2d Platoon, Company B, 112th Infantry, 28th Division, by Captain Phelps," HI, Box 2421, RG 407, NARA.

21 Questionnaire, Johnson.

22 Ibid.; Bradley, *Aid Man!*, p. 59.

23 E. G. Miller (1995) *A Dark and Bloody Ground: The Hürtgen Forest and the Roer River Dams, 1944–1945* (College Station: Texas A&M University Press), pp. 2–12; "Green Hell" J. B. McBurney (1999) *The 13th: A Private's Eye View* (Jennings, LA: John B. McBurney), p. 11; R. F. Weigley (1981) *Eisenhower's Lieutenants: The Campaign of France and Germany, 1944–1945* (Bloomington: Indiana University Press), p. 365; C. B. Currey (1984) *Follow Me and Die: The Destruction of an American Division in World War II* (New York: Stein and Day), p. 23.

24 Interview, "Report Hürtgen Forest Battle, 4th Division, First US Army, 7 November to 3 December 1944, by Captain K. W. Hechler, with the 1st Battalion, 22d Infantry, Division Narrative by Lieutenant Colonel William T. Gayle, 4th Division Information and Historical Service," Combat Interviews (CI) 34, Box 24021, RG 407, NARA.

25 Interviews, "Report Hürtgen Forest Battle, 4th Division by Hechler and Gayle," RG 407, NARA; "blow off" P. Boesch (1962), *Road to Huertgen: Forest in Hell* (Houston: Gulf Publishing), p. 164; McBurney, *The 13th*, p. 52.

26 McBurney, *The 13th*, p. 11; Miller, *A Dark and Bloody Ground*, p. 12; Weigley, *Eisenhower's Lieutenants*, 365; Interview, "2d Lieutenant George E Hackett, Battery A, 107th Field Artillery Battalion, 28th Division by Lieutenant Stockton, 55th General Hospital," HI, Box 24240, RG 407, NARA.

27 Questionnaire, Robert R. Reed II (28 April 2001).

28 Interviews, "Hürtgen Forest Replacements and Non-Battle Casualties, 1st Battalion, 22d Infantry, 4th Division, 16 November to 3 December 1944, with Captain Jennings Frye, S-1 1st Battalion, Lieutenant George Kozmetsky, Assistant Surgeon, 1st Battalion, Technical Sergeant 3rd Grade Harry I. Fingerroth, 1st Battalion Aid Station, Vicinity Gostingen, Luxembourg, 20 December 1944, Interviews by Captain K.W. Hechler, 2nd Information and Historical Services (VIII Corps)," CI 34, Folder II, Box 24021, RG 407, NARA. The fighting proved costly for all divisions that entered the forest, with the rifle platoons bearing the brunt of the losses. One medical soldier saw "scores of [dead] GIs lined up like cordwood...waiting for Graves Registration to collect their butchered and punctured bodies." Casualty statistics of two infantry divisions, the 28th and the 4th, expose the enormity of the mission that fell to company aid men and the medical soldiers at the BAS. The 28th Division suffered losses of over 6,000, while within a four-week period the 4th Division experienced a turnover approaching 200 percent in some rifle companies. These statistics reflect non-combat casualty evacuations including trench foot and pneumonia. The 28th Division report around 1,300 as non-combat casualties, while the 4th Division recorded 5,000 as combat killed or wounded, and 2,000 as non-combat. The 4th Division had already replaced rifle companies "almost twice over" during the first six weeks of Normandy, but had brought the division back to strength when it was ordered to the Hürtgen area. "scores" Paul Treatman as quoted in G. Astor (2000) *The Bloody Forest: Battle For Huertgen, September 1944–January 1945* (Novato, CA: Presidio Press), p. 197; "almost twice over," Interviews, "Report Hürtgen Forest Battle, 4th Division by Hechler and Gayle," RG 407, NARA; Weigley, *Eisenhower's Lieutenants*, p. 370; Ambrose, *Citizen Soldiers*, p. 170.

29 McBurney, *The 13th*, p. 11; Interviews, "Report Hürtgen Forest Battle, 4th Division by Hechler and Gayle," RG 407, NARA; Interviews, "Hürtgen Forest Replacements and Non-Battle Casualties, with Frye, et.al.," RG 407, NARA; "Interview, 2d Lieutenant Donald E. Martini, Company C, 110th Infantry, 28th Division by Lieutenant Posvar, 108th General Hospital," HI, Box 2421, RG 407, NARA.

30 See Chapter 5 concerning first echelon aid for non-combat casualties, including trench foot and combat exhaustion.

31 Interview, "Hürtgen Forest Campaign, 7 November–7 December 1944, with Captain Francis J. McCauley S-1, 2d Battalion, 12th Infantry, by Lieutenant Francis H. Fife," CI 34, Box 24021, RG 407, NARA.

32 United States 78th Division (1947) *Lightning: The Story of the 78th Infantry Division*, (Washington, DC: Infantry Journal Press), p. 63; Les Habegger, "A Medic," available from http://www.trailblazersww2.org/amedic.htm, Steve Dixon, editor, Internet, accessed 1 August 2002; A. N. Towne (2000) *Doctor Danger Forward: A World War II Memoir of a Combat Medical Aid Man, First Infantry Division* (Jefferson, NC: McFarland and Co., Inc.), p. 158.
33 Atwell, *Private*, pp. 137–172.
34 Habegger, "A Medic."
35 Atwell, *Private*, p. 162.
36 Ibid.
37 Questionnaire, Berry.
38 Questionnaire, Lovelace; B. W. Dohmann (1969), "A Medic in Normandy" *American History Illustrated*, (vol. 4, no. 3), p. 12.
39 "severe" *Report of the General Board*, "Study Number 92," p. 1, MHI; Cosmas and Cowdrey, *Medical Department*, p. 446.
40 Questionnaire, "never" Sullivan. See Chapter 4 concerning combat medics as a target of enemy fire.
41 Questionnaire, Joseph Cosby (19 May 2001). In contrast to Cosby, most infantry soldiers found themselves in the position of surrogate medics on a more limited basis. One rifleman recalled that at one point, an engineer came forward with them to clear the area of mines. The engineer accidentally "set one off and severed his leg just above the knee." The platoon medic was not with the unit at the moment, and two infantry soldiers immediately responded, using a belt as a tourniquet. Other combat soldiers report that they tended to their buddies on the occasions when the aid men were occupied with more critically wounded men, or even that German medics helped with their wounded on occasion. Questionnaires, "set one" James Hanson (31 March 2001), Wilfred B. Howsman, Jr. (September 2001), and Elvin Keen (10 August 2001); Interview, "Lieutenant Frederick Deutsch, Company I, 134th Infantry, 35th Division, by Lieutenant Stockton, 55th General Hospital," HI, Box 24240, RG 407, NARA; Interview, "Technical Sergeant Alex Fergusson, 346th Infantry, 87th Division," HI, Box 24240, RG 407, NARA; Interview, "2d Lieutenant John Hayduchok, Company C, 110th Infantry, 28th Division," HI, Box 2421, RG 407, NARA.
42 Questionnaires, Braunhardt and Pomplin; Bradley, *Aid Man!*, p. 51; Tsuchida, *Wear It*, p. 123.
43 Questionnaires, "until death" Allen, and Smith and Jacob E. Way (19 April 2001).

4
The Battalion Aid Station

Abstract: *This chapter analyzes the role of the BAS personnel during combat. In contrast to the aid men, the BAS medics could rely upon each other, working in concert in order to remove a wounded man from the company aid man's hands, carry him rearward, stabilize him, and evacuate him. To facilitate this teamwork, BAS medics also demonstrated uncommon versatility by assuming tasks for which they had not been trained, cooperating to move and aid the wounded.*

Shilcutt, Tracy. *Infantry Combat Medics in Europe, 1944–45*, Basingstoke: Palgrave Macmillan, 2013. DOI: 10.1057/9781137347695.

A soldier burst through the door of army doctor Frank Miller's tiny aid station yelling that his companion lay bleeding to death just outside. But Miller dismissed the pleading rifleman as a dozen other men lay bleeding to death in the station; the new casualty must simply be added to the gory line. Two days of house-to-house fighting through the once pastoral hamlet of Frenz, Germany had inflicted heavy losses on the 413th Infantry, including the 2d Battalion commander killed by an artillery burst as he stood a few feet from the battalion surgeon. Doc Miller and his BAS team hastily patched up the wounded who could be brought inside the bombed out basement, but the surgeon grew increasingly discouraged when the Germans blocked the supply and evacuation routes, shutting off hope for the wounded. The interdicted evacuation system meant Miller's team was losing men who could otherwise be saved.[1]

Frantic to move his wounded, Miller ordered his staff to load the most desperate casualties into an ambulance. The doctor jumped into the driver's seat and headed out of Frenz toward allied lines. Remarkably the Germans held their fire for the ambulance, allowing Miller and his human cargo to pass. This band of wounded soldiers owed their lives to Miller's refusal to yield, as in true combat medic practice, he circumvented procedure.[2]

Miller's action typifies the pragmatism of infantry BAS medics during the European war. Although the Medical Department mandated a detailed system for the expedient care and evacuation of the wounded, BAS medical men discovered that their fundamental goal of rapid evacuation often put them at odds with prescribed procedures. The theory for front line medical care held that a company aid man initially tended the wounded before teams of litter bearers transported the injured soldier to the BAS for interim treatment. Once in the station, protocol held that teams of medical technicians then worked under the direction of the battalion surgeon to treat the casualties before sending them back through the system for more definitive care.

But, like Miller's experience in Frenz, each combat encounter forced BAS personnel to reframe their plans for completing the task. Combat medical soldiers assigned to aid stations respected the intentions of the model but unhesitatingly acted creatively and practically to expedite treatment and evacuation.[3] Undaunted by combat's unpredictability, medics at the BAS, in common with company aid men, embraced function over structure to insure care for the wounded.

The physical burden of retrieving and hauling the wounded away from the battle line and rearward to the BAS fell to litter bearer teams working out of the station. One of the Army's most physically demanding jobs, moving casualties from one point to the next under combat conditions raised a series of challenges reaching beyond brute strength.[4] In ideal situations, medical corpsmen operated in teams of four to quickly locate the casualty, place him carefully onto the stretcher, and carry him back to a more stable environment for emergency care. But combat rarely provided the ideal situation.

Weather, terrain, and enemy fire compounded the litter bearer's journey through Europe. Bearers carried casualties day and night through summer's heat, relentless rain, slippery mud, and hip-deep snow drifts. They bore the wounded through icy streams, balanced stretchers as they traversed rocky gullies, and maneuvered through the tangled growth of forest and hedgerow. Under these most severe conditions litter bearers also dodged enemy fire making one-way hauls that sometimes exceeded two hours.[5]

Innovative litter teams often tossed aside protocol to ease the load and decrease the time spent on each haul back to the BAS. During the defense of the Vosges Mountains in December 1944, 63d Division litter bearers who encountered difficult terrain and climatic conditions faced evacuation "entirely different from the earlier teachings." Army procedure called for the use of jeeps to carry casualties, but the icy roads made it "hardly possible" for the motorized vehicles to reach the forward companies. In response, bearers outfitted sleds as litters, simplifying an arduous task.[6] Depending on terrain and weather, other medical soldiers mounted stretchers onto skis, or commandeered wheelbarrows and mules. And when circumstances allowed, combat medics co-opted motorized vehicles, particularly the versatile jeeps and weasels. One MAC officer pilfered a second jeep for his litter squads soon after his arrival at the BAS, calling his extra-legal appropriation the "best thing we ever did."[7]

Whether the litter bearers used mechanical equipment or physically carried the casualties, the ability to swiftly pinpoint the wounded depended upon timely communications. In the best of circumstances litter bearers waited far enough forward so that they received information directly from walking wounded or from line company radio operators. Aggressive litter squads serving the 116th Infantry carried headsets and tapped into the communication wires to hear first-hand the most current situation reports. But not all squads gained speedy or accurate information about the wounded. When communications collapsed, litter bearers systematically combed the combat zone for casualties, listening for cries from the company aid men or screams from the wounded. Certain markers, such as a rifle stuck in the ground upside down, might also alert the bearers.[8]

During engagements resulting in particularly heavy casualties, company aid men could not reach all the wounded since their instructions to stay with advancing troops meant they had to leave wounded behind. Litter bearers then assumed the responsibility for the initial treatment; in effect, while they were not assigned specifically as company aid men, all front line medics including litter bearers, functioned as aid men when necessary. When working as the first of the medical soldiers to reach the wounded, litter bearers stabilized the casualties before loading them onto their stretchers. Like company aid men, they carried aid pouches packed with emergency supplies to control bleeding, combat shock, and ease pain. Once they treated a wounded soldier, litter bearers started him on the journey rearward.[9]

Of course at every juncture, the most constant threat to the litter bearers came from German fire. One corpsman recalled the first time that his squad left the BAS to retrieve wounded. As they walked down a road toward the action, artillery boomed in the distance. The familiar sound strangely comforted the medic by reminding him of the field exercises in the states, but his ease was short-lived. When shells exploded onto the road ahead of him, he "dove for the ditch" while thoughts of disbelief flashed through his mind: war was not like training.[10]

The medics' job demanded that they move directly into the combat zone even as fighting units dug in, earning some litter squads a reputation for recklessness. Battalion surgeons often urged them to wait for safer circumstances or to abandon altogether some extremely dangerous attempts at rescue, for casualty statistics had the potential to multiply each time a squad approached an injured man as the battle still raged.[11] Even after safely rescuing a casualty from the field, litter bearers often remained under enemy fire as they returned to the BAS.

One four-man litter squad working with the 30th Division in July 1944 had retrieved a casualty near St. Lô, and as they hiked the wounded soldier toward the BAS, a German tank shell blew up on the road just ahead of them. The medics dropped the stretcher, their patient crashed to the ground, and the bearers buckled from the explosion. When the firing ceased, one of the litter bearers, 19-year-old Private Everett Smith, clambered to his feet, all the while checking his body for shrapnel wounds. Stunned that he was physically unharmed, Smith pulled at his aid kit strap and realized that the flying debris had completely shredded his pouches. He then turned to assist his fellow medics who had sustained serious wounds and the casualty who had been further injured—all this on the bearer's first day in battle.[12]

These of the 30th Division, as all front line medical soldiers, worked under the pressure of trying to evade impersonal shrapnel, yet at times they came face to face with the enemy and even labored unharmed during brief and informal truces.[13] Additionally, if the Germans recognized the soldiers as non-combatant medical men, they usually held their fire, allowing the medics to return unmolested to their lines with the wounded.[14]

When fighting opened in Europe, medics had worn standard-issue brassards, white bands with red crosses, on their left arms as a means of distinguishing themselves from the fighting troops. Yet some medics felt the white bands, often bloodied and dirtied, were not conspicuous enough to be seen and sought ways to signify more clearly their noncombatant status. Aid men and litter bearers began painting their helmets with large red crosses framed by round fields of white, speculating this provided a heightened level of protection.[15] One division reported that their "medical aid men had everything to gain and nothing to lose by painting their helmets," because experience showed that when litter bearers with marked helmets appeared on the field enemy fire ceased.[16]

Medics met with enough success when they made the changes to the helmets that by November 1944, the Army endorsed and standardized the practice. But the heightened identification did not always ensure safety.[17] One 4th Division motorized retrieval squad in the neighborhood of Villedieux les Poel went beyond simple helmet markings and also painted large crosses on their vehicles and prominently flew a white flag bearing a red cross from their jeep. Despite these efforts, the Germans opened fire on the American medical jeep as it rounded a curve in the road, but upon realizing what they had done, the Germans ceased firing and even apologized. One medic was lost in the incident.[18]

The frequent deaths of front line medics later compelled some medical soldiers to question the purported advantage of the brassards and marked helmets. Many medical soldiers worried that, rather than providing protection, the Red Cross insignia made an inviting target, so some combat medics stripped the brassards and scraped the paint from their helmets just prior to engagement.[19] Occasionally rumors filtered back to the BAS that medical soldiers were being "taken out," and confirmation came when litter bearers retrieving the bodies of slain medical buddies found that the fatal bullets had pierced the helmet.[20] Whether as a result of intentional or random fire, medics did at times receive serious head injuries. In late November 1944 Private William A. Reed of the 1st Battalion, 22d Infantry aid station led a litter team into the woods west of Grosshau when shrapnel pierced his helmet. Insisting that he was not hurt Reed made two additional litter hauls of over 1,000 yards before collapsing. When the surgeon examined Reed he discovered that the litter bearer had continued working under extreme combat conditions with a fractured skull.[21]

When heavy fire or German control of the combat zone impeded attempts to reach casualties medical soldiers sometimes called on the infantry for help.[22] Rifle companies "cleaned out" forward positions to facilitate evacuation, laid protective cover as litter bearers worked, and occasionally accompanied litter teams into occupied territory.[23] Most often, litter squads operated without direct protection, moving back and forth from the battlefield to the BAS as the battle surged around them. In one costly encounter in January 1945, eight litter bearers of the 35th Infantry Division remarkably evacuated 61 men over the course of two days. Despite heavy artillery and mortar fire that kept the infantry pinned down, the litter bearers repeatedly moved into the fire zone, crossing a footbridge to reach the casualties.[24]

The extraordinary conditions imposed by terrain, weather, and enemy fire meant that litter bearers periodically needed reinforcement, and they unhesitatingly impressed German prisoners as well as non-medical American soldiers to help carry wounded to the rear. The devastation of the Hürtgen Forest fighting led BAS personnel to extreme measures; they conscripted cooks, Ammunition and Pioneer (A&P) personnel, and administrative soldiers to augment the litter squads. For one A&P platoon, the move to the front proved costly; 11 A&P men lay among the casualties, alongside the wounded they had gone to help. In the 100th Division the need for assistance meant the impressments of the unit's band, men completely unaccustomed to the realities of combat medical duties. During engagements with unusually heavy losses, non-medical soldiers like these served as temporary replacements on litter squads, but most often rear echelon medical soldiers simply moved to the front, acting as supplemental or temporary frontline litter bearers.[25]

As the war in Europe evolved, heavier use of motorized evacuation between the regimental and divisional stations at times freed rear echelon litter bearer squads to work as temporary replacements for combat litter bearers or aid men. As soon as they completed their tasks these rear echelon soldiers pulled back to their original units, usually located several miles behind the line. Because the combat zone fluctuated some rear echelon medics had worked under fire, but most were unfamiliar with forward battle conditions. Yet these rear echelon litter bearers carried a strong commitment to front line objectives as they worked under heavy fire and without rest to help clear the fields and move the casualties to the station.[26]

One 70th Division BAS action typifies the gruesome nature of the scenes repeated in stations across the European continent. The medical team in the station sprang into action as litter bearers carried a casualty into the BAS, his torso ripped open, exposing the "complete intestinal system as it floated in a bath of blood."[27] Even as one team worked on the gaping wound, two more litter teams delivered additional casualties, each with a blood-soaked dressing on his right leg marking the place where a foot should have been. As medical technicians moved to the sides of the stretchers to release the tourniquets, the wounded men told how one had stepped on a mine and that his buddy triggered another while trying to help. The medics changed the bandages, shifted the casualties to a waiting ambulance, and as the vehicle drove away the medical soldiers rushed back to the aid room to handle the next casualty. There was little rest from "the tension of seeing these mangled pieces of flesh 24 hours a day."[28]

Under a primary directive to remain forward with the fighting elements, the entire BAS staff helped move and set up the aid station, frequently shifting multiple times a day.[29] Wide-ranging combat conditions dictated the location and establishment of the BAS, and medical soldiers adapted to combat conditions in order to achieve their goals. First and most important, since the BAS did not function as a definitive treatment center, all activity focused on "maintaining status quo" of the wounded by stabilization.[30]

BAS medics directed care toward impeding any further deterioration of the casualty's condition, so that the wounded man could make the trip to the rear successfully. They accomplished this principally by following up on the initial care provided by the front line aid man: limiting blood loss with bandages and tourniquets, controlling shock with plasma, relieving pain with morphine, limiting infection with sulfanilamide, and immobilizing fractures and breaks.[31] The transient nature of the BAS meant that there might not be room for more than six litters at a time, so prompt attention and evacuation had to free treatment spaces for more casualties.[32]

BAS medics usually tried to anticipate a generalized area where casualties might be concentrated and established their stations as far forward as was safe and workable. But environmental conditions along with the pace of combat usually proved the most influential determinants not only of the BAS location but also the form, which the station assumed. Dependent upon circumstances, the BAS might be situated as close as 100 yards from the fighting or as far as three miles behind the combat line. Available physical resources dictated the site of the stations, and medical personnel commandeered a variety of accommodations, ranging from existing structures to makeshift facilities.[33]

Those combat zones that offered inherent protection allowed BAS personnel to move their stations close to the line. In the French hedgerows shortly after D-Day, medics immediately recognized the folly of the training practice of setting up tents in open fields. Instead, they created mobile stations, moving in close proximity to the earthen barriers, treating casualties out of the back of transport and supply vehicles on the rapid advance.[34] As medics pushed forward with the fighting units through towns and villages, house-to-house combat allowed a BAS to remain somewhat protected, but close to the fight, as they converted intact or even partially destroyed buildings into aid stations.[35]

The efforts to evacuate the wounded rapidly at times assumed a measured pace, but during periods of heavy casualties the battalion surgeon treated the gravest injuries immediately, tasking his staff with handling those suffering from less critical wounds, confident in their abilities, even though they were not physicians. While surgeons dealt with gaping wounds, compound fractures, and casualties with multiple serious injuries, the medical technicians redressed bandages, administered plasma or morphine, applied splints, and sutured less serious wounds. Other medical soldiers followed through by handling the ever-present paperwork, but even as the medics carried on, the war at times filtered into the station with gunfire tearing through the BAS as litter bearers tried to move casualties in and out. The noise level within the station intensified when the crash of ordnance mingled with the groans and shouts of those reviving following transfusions; plasma carried a promise of life, but it also restored sensation to once-numbed bodies. Despite the din and the commotion, medical soldiers pushed on.[36]

Battalion surgeons prioritized the order in which casualties left for the rear.[37] They sent first those "most urgently in need of higher level care," but one surgeon recalled that since the broad mission was to preserve the fighting strength, he more than once made the "dreadful decision [to evacuate] not the man with the most serious wound, but the man who potentially could return to combat duty" most swiftly.[38]

The urgency of remaining near the combat units occasionally drove some medics to abandon the structural confines of a prescribed station altogether; in essence the BAS was not a physical entity, rather it was the medics and the care they gave. BAS personnel of the 88th Division traveled in Italy with battalion headquarters and set up their aid station only when headquarters stopped, generally when the unit stood down in reserve. But on the advance, the medical soldiers simply joined the line soldiers, carrying multiple aid kits jammed with supplies so that they could treat the wounded as quickly as possible. Most of the wounded men whom they tended never saw the battalion surgeon nor processed through any BAS; instead ambulances, or sometimes mules, took them directly to rear echelon facilities. If fighting blocked the evacuation route, then a litter bearer remained behind with a group of the wounded, not to man any sort of formal aid station, but only to wait until transportation arrived to pick up the wounded. The litter bearer then rejoined the fighting troops.[39]

While the medical men of the 88th Division stations adapted to the demands of warfare by contriving temporary stages and dispensing with setting up a BAS altogether, more commonly across Europe medical personnel set up collecting points to facilitate prompt evacuation to an established BAS. If the BAS could not get well forward, or if the line remained stationary, litter squads sometimes established collecting stations close to the company command post (CP). The communications network between the CP and the line companies meant litter bearers could locate the wounded and haul them back to the collecting station, where, terrain allowing, a jeep waited to take them to the BAS. This freed the litter bearers to return to the field and retrieve more wounded. While the collecting points permitted front line medical personnel to reach their objectives more effectively, other BAS medics who could not move the station close to the line adopted a more radical plan for prompt evacuation by establishing forward aid stations (FAS).[40] Combat conditions marked by rapid advance, excessive casualties, complicated evacuation routes, or a wide front presented a dilemma for BAS personnel who had set up miles behind the line. On these occasions it was impractical to dismantle and move the BAS, but the long litter hauls to the station meant delayed treatment and increased losses. Front line medical personnel coped with these difficulties by modifying standard procedure and splitting the station. More than merely multiplying collecting points for the wounded, the division usually resulted in the creation of a makeshift forward aid station operating within a few hundred yards of the line companies while retaining the more stable BAS to the rear.[41]

The FAS functioned as an intermediate medical resource, providing stopgap treatment between the initial ministrations of company aid men and the fuller care offered in the BAS. If conditions allowed, jeeps hauled rudimentary equipment forward; if not, medical or infantry soldiers packed in the medical gear. Most often the non-physician MAC officers or NCOs directed medical technicians and litter squads who worked out of an interim station, sometimes little more than a shelter half floating between trees. Despite the provisional nature of the FAS, the proximity to the line meant significantly shorter litter hauls and more immediate critical aid. The innovation of the forward stations meant that greater numbers of wounded received prompt attention and speedier evacuation than would have been possible with a unified BAS far behind the line.[42] Once again, medical soldiers improvised, ignoring regulation structures in order to heighten their effectiveness.

Combat medics also adjusted, even if sometimes reluctantly, to a variety of jobs in order to expedite treatment.[43] With a BAS handling up to 75 casualties a day during heavy engagements, it was important that each medic do "whatever needed to be done" to preserve the fighting strength of the unit.[44] During their introduction to combat emergency care, BAS personnel ordinarily experienced a brief period of confusion, but they soon "settled into an efficient routine [with each] picking up wherever the load required an extra hand."[45] Battalion surgeons did sometimes leave the confines of the station to act as litter bearers or to render aid closer to the front, but they more commonly kept to the BAS during combat. This meant that junior officers and enlisted men assumed yet more roles covered inadequately—if at all—in their stateside training.

Apart from staffing the forward aid stations, MAC officers, NCOs, and medical technicians joined litter teams when they were "shorthanded"; at the same time litter bearers substituted for company aid men if called on or if a line company lacked a medic.[46] One company aid man who occasionally did duty in the BAS recalled that he always felt safer there. Litter bearers echoed his sentiments because they thought that a move to the line companies guaranteed their death. Yet they "filled in wherever necessary" to advance the team's effort, even though they preferred the limited safety offered by the BAS.[47]

Along with the teamwork needed to treat the casualties, BAS medics also worked in concert to complete requisite paperwork that accompanied wounded soldiers moving to the rear. Emergency medical tags constituted the primary mechanism for essential information, but in the frenetic violence of the front lines, company aid men frequently abandoned this procedure, so BAS medics recorded crucial information on this mini-chart. The EMT identified the soldier, but more significantly for rear echelon medical personnel, it documented tourniquet application time, the amount of morphine administered, and other vital treatments that had been given. At times BAS medics, in an effort to speed the process of filling out EMTs, simply stamped the wounded man's dog tags onto the form.[48]

On occasion BAS personnel found release from the pressures of emergency care, even while under fire. For some, music proved soothing, and somehow, even in the swirl of war, accordion music and organ tunes occasionally filtered through the battle noises.[49] Other units found that humor dispelled tense situations. One battalion surgeon recalled that his immediate supervisor, a regimental surgeon, always phoned before visiting the BAS if the combat unit was in contact with the enemy. The BAS medics picked up on a pattern developing with these telephone conferences. If there was an absence of fire during "idle talk," the superior officer would appear at the BAS within a short time, but if any shelling could be heard through the phone lines, the regimental surgeon showed up "substantially later in the day" or not at all. The enlisted men turned the episodes into profitable ventures, establishing a betting pool on when or whether the surgeon would appear.[50]

As American soldiers fought the war in Europe through hedgerow country, in forested areas, across mountainous regions, and into villages and towns, medics of the battalion aid stations lent all possible aid to wounded soldiers, whom one army surgeon called "instruments singing pain."[51] In ordinary circumstances BAS soldiers enjoyed a degree of safety denied company aid men, yet combat often shrugged off the ordinary. Medical technicians assigned to the more rearward BAS sometimes found themselves pressed into duties alongside or even in place of company aid men, and they unfailingly performed these more hazardous tasks with dispatch, courage, and effectiveness.

Whether in such unanticipated situations or tied more closely to the regulation posts, BAS medical men showed the same adaptability and inventiveness which marked the aid men who labored on the combat line. They quickly modified or even abandoned the BAS structures learned in stateside training centers, following their cardinal duty of providing the best aid possible for the wounded. And since infantry losses represented more than 85 percent of the war's total casualties, with some units turning over two and three times during their European service, the effective care provided by the BAS medics, after the initial treatment by the company aid men, proved critical to the survival of the wounded who fell into their devoted hands.[52] From harried battalion surgeons to the hard-driving litter bearers, the medical soldiers of the BAS joined the company aid men to complete the first link in the chain of life for the infantry soldiers whose war this was.

Notes

1 Frank L. Miller to author, 16 March 2001, in author's possession; Questionnaire, Karl Stelljes (24 May 2001); Karl Stelljes to author, 30 May 2001, in author's possession.
2 Questionnaire, Stelljes; Stelljes to author, 30 May 2001.
3 Medical officers directed the activities of and commanded the soldiers assigned to the aid stations. The highest ranking officer, usually a captain, was the battalion surgeon, a medical doctor. As the war evolved, many of the administrative duties inherent in the function of the BAS fell to the assistant battalion surgeon, usually a MAC officer. Non-commissioned officers assumed clerical and liaison duties and most often took charge of the litter squads. Medical technicians helped the surgeon with emergency diagnosis and immediate treatment of combat wounded. Questionnaires, Carl Aschoff (14 June 2001), Stephen J. Barnett (28 June 2001), Frank R. Ellis (4 June 2001), David E. Fought (17 April 2001), Allen L. Johnson (n.d.), Buster M. Simmons (8 January 2002), and Donald Warner (16 April 2001); Graham A. Cosmas and Albert E. Cowdrey (1992) *The Medical Department: Medical Service in the*

European Theater of Operations (Washington, DC: Center of Military History), p. 368; also see note 30.

4 Questionnaires, Aschoff, Ellis, Johnson, and Neel Price (5 June 2005); "Battle Experiences, European Theater of Operations United States Army, Extracts, July 1944–45," Number 48, 31 January 1945 and Number 74, 5 March 1945, Box 24148, RG 407, NARA; "After Action Report, 13 July to 1 August 1944, 5th Medical Battalion, by Captain Charles F. Byrd, Medical Administrative Corps," Box 6919, RG 407, NARA; "Unit Report Number 2, 1 August 1944 to 31 August 1944, 103d Medical Battalion, by Major Edgar W. Meiser, Medical Corps, 103d Medical Battalion Acting Commander," Box 8621, RG 407, NARA; L. Atwell (1958) *Private*, (New York: Simon and Schuster), pp. 56–154; W. S. Boice (1959) *A History of the Twenty-Second United States Infantry in World War II* (np), pp. 61–64.

5 Questionnaires, "endless" Aschoff, and Ellis, Johnson, and Warner; S. Dixon, ed. "70th Division Association Website". L. Habegger, "A Medic" Available http://www.trailblazersww2.org/amedic.htm (accessed 1 August 2002); Interviews, "4th Division, Hürtgen Forest Battle, 7 November to 11 December 1944," CI Number 34, Box 24021, RG 407, NARA; "Unit History, 63d Infantry Division, Office of the Surgeon, 30 June 1945," Box 11162, RG 94, NARA; Interview, "Battle of Hürtgen Forest, 16 November–3 December 1944, Battalion Aid Men 1st Battalion, 22d Infantry Regiment, 4th Division, with Lieutenant George Kozmetsky, Assistant Surgeon, 1st Battalion, Technical Sergeant, Third Grade Harry I. Fingerroth, 1st Battalion Aid Station, Technical Sergeant, Fifth Grade Joseph J. Thomas, Aid Man with Company B, Technical Sergeant, Fifth Grade Wade H. Carpenter, Aid Man with Company B, by Captain K. W. Hechler, 20 December 1944," CI, Box 24021, RG 407, NARA; Interview, "Report Hürtgen Forest Battle, 4th Division, First US Army, 7 November to 3 December 1944, by Captain K.W. Hechler, with the 1st Battalion, 22d Infantry, Division Narrative by Lieutenant Colonel William T. Gayle, 4th Division Information and Historical Service," CI 34, Box 24021, RG 407, NARA; Interview, "Lieutenant Benjamin W. Mills, Commanding Officer, and 1st Sergeant Lawrence E. Houck, Company F, 8th Infantry, 4th Division, 18 August 1944," Box 24021, RG 407, NARA; Interview, "Hürtgen Forest Campaign, 7 November–7 December 1944, with Captain Francis J. McCauley S-1, 2d Battalion, 12th Infantry, by Lieutenant Francis H. Fife," CI 34, Box 24021, RG 407, NARA; Interview, "Lieutenant Colonel Kenneth R. Lindner, Commanding Officer, 3rd Battalion, Major Herman R. Rice, Jr., Executive Officer, 3d Battalion, 1st Lieutenant Cornelius R. O'Donnell, et al., 16–24 December 1944, 3d Battalion, 12th Infantry (right flank) 4th Division, Battle of Luxembourg, Osweiler and Dickweiler, by Lieutenant Colonel William T. Gayle, Captain D. G. Dayton, and Lieutenant S. J. Tobin," 27 December 1944, Box 24022, RG 407, NARA; 78th Division,

Lightning, p. 45; Atwell, *Private*, pp. 56–154; Boice, *A History of the 22nd*, pp. 61–64; "Battle Experiences, Extracts, July 1944–1945," RG 407, NARA; "Battle Experiences," Number 48, 31 January 1945 and Number 74, 5 March 1945, RG 407, NARA; "After Action Report, 13 July to 1 August 1944, 5th Medical Battalion, by Captain Charles F. Byrd, Medical Administrative Corps," Box 6919, RG 407, NARA; "Unit Report Number 2, 1–31 August 1944, 103d Medical Battalion, by Major Edgar W. Meiser, Medical Corps, 103d Medical Battalion Acting Commander," Box 8621, RG 407, NARA.

6 "Unit History, 63d Division," RG 94, NARA.

7 Weasels were small, tracked vehicles. "Unit History, 63d Division," RG 94, NARA; Questionnaires, "best thing" Aschoff, and T. William Bossidy (1 July 2001), Price, and Everett Smith (7 August 2001); Interview, "McCauley by Fife," RG 407, NARA; Interview, "Lieutenant Charles R. Crispin, 23d Infantry, 5th Division," HI, Box 24240, RG 407, NARA; Interview, "Lieutenant August T. McColgan, 26th Infantry, 1st Division," HI, Box 24240, RG 407, NARA; Interview, "Private Harry D. Moffatt, Company G, 120th Infantry, 30th Division, 192d General Hospital," HI, Box 24240, RG 407, NARA; Joe Hochadel to author, 9 November 2001; "Periodic Report, 8th Medical Battalion, 1 January 1945 to 1 June 1945, account describing isolation of the 3d Battalion, 121st Infantry by Major Anthony S. Terranova, Medical Corps, 121st Infantry, 8th Division," Box 7322, RG 407, NARA; D. Pergrin with E. Hammel (1989) *First Across the Rhine: The 291st Engineer Combat Battalion in France, Belgium, and Germany* (New York: Antheneum), p. 23.

8 Questionnaires, Frank J. Irgang (March 2001) and Walter D. Murchison (8 May 2001); "Battle Experiences" Number 74, 5 March 1945, RG 407, NARA; "Unit History, 63d Division," RG 94, NARA; Interviews, "Battle of Hürtgen Forest, with Kozmetsky, et.al.," NARA; G. Wilson (1987) *If You Survive* (New York: Ivy Books), p. 151.

9 Questionnaires, Roy Barratt (13 June 2001) and Smith; Atwell, *Private*, 172; Boice, *A History of the 22nd*, pp. 61–64.

10 P. W. Sewell, ed. (2001) *Healers in World War II: Oral Histories of Medical Corps Personnel* (Jefferson, NC: McFarland and Company), p. 6.

11 Questionnaire, "reckless" Ellis; Interview, "Captain Elmer G. Koehler, Medical Detachment, 2d Battalion, Battalion Surgeon, 12th Infantry, by Captain Phelps, 111th General Hospital," HI, Box 24242, RG 407, NARA; Interviews, "Officers, 2nd Battalion, 12th Infantry, 4th Division," 7 August 1944, Folder 31, Box 24021, RG 407, NARA; "General Orders Number 8, 22 January 1945, After Action Report 110th Medical Battalion, January to June 1945, 35th Division," Box 9790, RG 407, NARA.

12 Questionnaire, Smith.

13 Questionnaires, Kenneth T. Delaney (n.d.) and Johnson; C. D. Curley, Jr. (1998) *How a Ninety-day Wonder Survived the War: The Story of a Rifle Platoon*

Leader in the Second Indianhead Division During World War II (Richmond, VA: Ashcroft Enterprises), p. 65; B. W. Dohmann (1969), "A Medic in Normandy" *American History Illustrated*, (vol. 4, no. 3), pp. 12–13; M. Griesbach, ed. (1988) *Combat History of the Eighth Infantry Division in World War II* (Nashville: Battery Press), p. 23; G. S. Johns (1958) *The Clay Pigeons of St. Lô* (Harrisburg, PA: Military Service Publishing Co.), p. 93; Interview, "Captain David M. Hull, Company M, 3rd Battalion, 110th Infantry, 28th Division, 155th General Hospital," HI, Box 2421, RG 407, NARA.

14 Questionnaires, Barnett and Elvin Keen (10 August 2001); Griesbach, *Combat History*, 23; Dohmann, "A Medic in Normandy," pp. 12–13; "Report of Action Against the Enemy, 4 October 1944, 110th Medical Battalion, 35th Division," Box 9790, RG 407, NARA; Johns, *The Clay Pigeons of St. Lô*, p. 24; H. P. Leinbaugh and J. D. Campbell (1985) *The Men of Company K: The Autobiography of a World War II Rifle Company* (New York: William Morrow Company), p. 212; Wilson, *If You Survive*, p. 31.

15 Questionnaires, Donald F. Eberhart (10 August 2001), Serge Manni (25 September 2001), and John T. Sullivan (n.d.); "Medic Medic," *The Trailblazer*, a publication of the 70th Infantry Division Association, (Winter 1998): pp. 6–9; "Report of Operations, September 1944, 180th Infantry, 45th Division, Transmittal of Organization History, 15 October 1944," Box 11098, RG 407, NARA; R. L. Sanner (1995) *Combat Medic Memoirs: Personal World War II Writings and Pictures* (Clemson, SC: Rennas Productions), p. 641.

16 "Report of Action Against the Enemy, 110th Medical Battalion, 4 October 1944," RG 407, NARA.

17 Questionnaires, Barnett and John Rheney, Jr. (10 March 2001); Cosmas and Cowdrey (1992) *Medical Department*, p. 372; R. B. Bradley (1970) *Aid Man!* (New York: Robert Bradley), p. 49.

18 Interviews, "Koehler by Phelps" and "Officers," RG 407, NARA.

19 Questionnaires, Barnett, William Braunhardt (30 May 2001), and Keen; Griesbach, *Combat History*, 23; Dohmann, "A Medic in Normandy," pp. 12–13; "Report of Action against the Enemy, 4 October 1944, 110th Medical Battalion, 35th Division," Box 9790, RG 407, NARA; Johns, *Clay Pigeons*, p. 24; Leinbaugh and Campbell, *Men of Company K*, pp. 212; Wilson, *If You Survive*, p. 31.

20 "taken out," 70th Infantry Division Association "Medic Medic," *The Trailblazer* (Winter 1998), p. 6; "Report of Operations, September 1944, 180th Infantry, 45th Division," RG 407, NARA; Habegger, "A Medic"; P. Boesch (1962) *Road to Huertgen: Forest in Hell* (Houston: Gulf Publishing), p. 9.

21 Interviews, "Battle of Hürtgen Forest, with Kozmetsky, et al." and "Officers," RG 407, NARA.

22 Questionnaire, Aschoff; Interview, "McCauley by Fife," RG 407, NARA; Boice, *A History of the 22nd*, 61–64; Boesch, *Road to Huertgen*, p. 241.

23 Interview, "cleaned out" "McCauley by Fife," RG NARA; 70th, "Medic! Medic!," pp. 6–9; Habegger, "A Medic"; Questionnaire, Ellis.
24 "General Orders Number 8, 22 January 1945, After Action Report 110th Medical Battalion, January to June, 1945, 35th Division," RG 407, NARA.
25 Questionnaires, Gerald W. Allen (25 March 2001), Aschoff, Wilbur Heinold (2 April 2001), Keen, Murchison, Russell Wade Redfern (1 June 2001), and Smith; Boice, *A History of the 22nd*, pp. 61–64; Cosmas and Cowdrey, *Medical Department*, pp. 364–372; Interview, "Lieutenant Robert T. Corns, Company H, 358th Infantry, 90th Division," HI, Box 24240, RG 407, NARA; "General Orders Number 5, 14 January 1945, After Action Report 110th Medical Battalion, January to June, 1945, 35th Division," Box 9790, RG 407, NARA; Interviews, "McCauley by Fife" and "Lindner, et.al. by Gayle, et. al.," RG 407, NARA; "Battle Experiences," Number 16, 20 December 1944, RG 407, NARA; Interviews, "Hürtgen Forest Battle, 7 November to 11 December 1944, 4th Information and Historical Service, Ninth United States Army, 3 May 1945," CI, Box 24201, RG 407, NARA; C. B. Currey (1984) *Follow Me and Die: The Destruction of an American Division in World War II* (New York: Stein and Day), p. 98; V. M. Lockhart (1981) *T Patch to Victory* (Canyon, TX: Staked Plains Press), p. 47.
26 Each infantry division's organic Medical Battalion had three collecting companies (A, B, C) and one clearing company (D). Every collecting company received assignment as part of the combat team, one per infantry regiment. Collecting companies planned for and executed evacuation from the BAS to the second echelon facilities but on occasion, collecting company litter bearers transferred temporarily to first echelon facilities to help with evacuations from the battle field to the BAS. Rear echelon medical soldiers reassigned to the BAS came under the command of the battalion surgeon and while this command shift best facilitated the evacuation process, on occasion it prompted friction between the battalion surgeon and the collecting company commander. Rear echelon litter bearers who felt they were being used needlessly sometimes by-passed the battalion surgeon, and requested through their original commanders that they be called to work on the front only if BAS litter bearers' numbers proved insufficient. When a group on temporary assignment felt it was being sacrificed to save BAS personnel, morale among the temporaries at the station dropped dramatically. Despite such occasional disputes over authority, most BAS personnel, both temporary and permanent, worked together successfully to remove the wounded from the field. "Periodic Report, 8th Medical Battalion, 1 January to 1 June 1945, Annual History," pp. 3–34, Box 7322, RG 407, NARA; "History of the 8th Medical Battalion, 11 February 1945, 8th Division," Box 7322, RG 407, NARA; 78th "Medic! Medic!"; A. N. Towne (2000) *Doctor Danger Forward: A World War II Memoir of a Combat Medical Aid Man, First*

Infantry Division. (Jefferson, NC: McFarland and Co., Inc.), pp. 116, 151; Cosmas and Cowdrey, *Medical Department*, pp. 364–372; Questionnaires, Jesus (Jesse) Armendariz Sr. (9 September 2001), Howard T. East, Jr. (n.d.), and Walker Powe (16 August 2001); "After Action Report by Byrd" 5th Medical Battalion, RG 407, NARA; G. Astor (2000) *The Bloody Forest: Battle for Hürtgen, September 1944–January 1945* (Novato, Ca: Presidio Press), pp. 107, 197; "Unit Report of Operations, 3d Medical Battalion, 3d Division, 1–30 December 1944, Operational Reports," Box 6246, RG 407, NARA; "After Action Report, December 1944, 5th Medical Battalion, 5th Division," Box 6919, RG 407, NARA; "Resume of Sanitary Report for November, 1 December 1944, Medical Department Problems and Activity, 44th Division," Box 10804, RG 407, NARA; "History of Company C, 1st Medical Battalion, 1st Infantry Division," Box 5967, RG 407, NARA; "Unit Report Number 3, 103d Medical Battalion, 28th Division 1 October 1944," Box 8621, RG 407, NARA.
27 78th, "Medic! Medic!," 9.
28 Ibid.
29 MAC officers became increasingly important in the late summer of 1944, taking the place usually of an MC officer; they relieved medical doctors of the administrative duties of the station, but also helped as needed with medical tasks; see Chapter 1 for additional details on training of these non-medically educated officers. Questionnaires, Aschoff, Barnett, Ellis, Fought, Johnson, Simmons, and Warner; Cosmas and Cowdrey, *Medical Department*, p. 368; R. R. Taylor, W. S. Mullins, R. J. Parks (1974) *Medical Training in World War II* (Washington, DC: Office of the Surgeon General).
30 Questionnaire, "maintaining" Brown McDonald, Jr. (22 June 2001); M. Fishbein, ed. (1945) *Doctors at War* (New York: EP Dutton and Co.), p. 120.
31 Questionnaires, Aschoff, Frank Miller (2 April 2001), Powe, and Warner; J. Greenberg and H. C. McKeever, eds (1995) *Letters From a World War II GI* (New York: Franklin Watts), pp. 59–89; Interview, "McCauley by Fife," RG 407, NARA.
32 Questionnaires, Aschoff, Miller, Powe, and Warner; Greenberg and McKeever, *Letters*, pp. 59–89; Interview, "McCauley by Fife," RG 407, NARA.
33 "Battle Experiences," Number 18, 12 August 1944 and Number 74, 5 March 1945," RG 407, NARA; Questionnaires, Allen, Aschoff, Bossidy, Ellis, Miller, Powe, Price, and Warner; Cosmas and Cowdrey, *Medical Department*, pp. 364–372; Sanner, *Memoirs*, pp. 59–89; Currey, *Follow Me and Die*, p. 137; Greenberg and McKeever, *Letters*, pp. 55–90; 78th Division, *Lightning*, p. 6; Astor, *Bloody Forest*, p. 197.
34 Questionnaires, Johnson and Smith; F. J. Irgang (1949) *Etched in Purple* (Caldwell, OH: The Caxton Printers, Ltd.), pp. 28–29; Cosmas and Cowdrey, *Medical Department*, p. 293.

35 Questionnaires, Aschoff, Ellis, Murchison, and Smith.
36 Questionnaires, Aschoff, Ellis, Keen, McDonald, Miller, Powe, Price, and Warner; Atwell, *Private*, p. 172; Interview, "McCauley by Fife, RG 407, NARA; Greenberg and McKeever, *Letters*, pp. 59–89; 78th, "Medic! Medic!," pp. 6–9; J. B. Mittleman (1948) *Eight Stars to Victory*, (Washington, DC: 9th Infantry Division Association).
37 Questionnaires, Ellis, McDonald, Miller, Powe, and Warner.
38 Questionnaires, "most urgently" Ellis, and "dreadful decision" Powe.
39 Questionnaires, Barratt and Price.
40 "radical," "Battle Experiences," Number 74, 5 March 1945, RG 407, NARA; "Battle Experiences, July 1944–45," p. 64, RG 407, NARA; Questionnaires, Ellis, Glen J. Mitchell (22 August 2001), and Redfern; Atwell, *Private*, p. 109.
41 Questionnaires, McDonald and Warner; "Battle Experiences," Number 48, 31 January 1945 and Number 74, 5 March 1945, RG 407, NARA; "Unit History, 63d Division," RG 94, NARA; Interviews, "Battle of Hürtgen Forest, with Kozmetsky, et.al.," RG 407, NARA; Cosmas and Cowdrey, *Medical Department*, pp. 366; Greenberg and McKeever, *Letters*, pp. 59–89.
42 Questionnaires, Barnett, Bossidy, Ellis, Fought, Miller, Murchison, Redfern, and Smith; "Unit History, 63d Division," RG 94, NARA; Interviews, "Battle of Hürtgen Forest with Kozmetsky, et.al.," RG 407, NARA; "Periodic Report 8th Medical Battalion, 1 January 1945–1 June 1945, account of capture and subsequent release by the German Army by Captain Raymond D. Markle, Medical Corps, 7 September 1944," Box 7322, RG 407, NARA; Cosmas and Cowdrey, *Medical Department*, p. 366; Greenberg and McKeever, *Letters*, pp. 59–89.
43 Questionnaires, Armendariz, Aschoff, Barnett, Bossidy, Eberhart, Ellis, Keen, Miller, Murchison, Powe, Smith, and Warner; Greenberg and McKeever, *Letters*, pp. 59–89.
44 Questionnaires, "whatever" Murchison, and Powe; Currey, *Follow Me and Die*, p. 92.
45 Questionnaire, Ellis.
46 Former medics G. Allen, Berry, and Murchison report their commitment to the unit was such that they either refused the opportunity to be rotated with litter bearers, or they returned immediately to their group after recuperating from wounds. Questionnaires, "shorthanded" Aschoff, and G. Allen, Barnett, Francis Irvin Berry (29 May 2001), Bossidy, Eberhart, McDonald, Murchison Miller, Powe, and Warner.
47 Questionnaires, Barnett and Johnson; Atwell, *Private*.
48 Questionnaires, Ellis, Miller, and Warner.
49 Questionnaire, "relieved" Ellis; "one more" W. S. Tsuchida (1947) *Wear it Proudly: Letters* (Berkeley: University of California Press), pp. 26–27.
50 Questionnaire, Ellis.

51 B. Phipps (1987) *The Other Side of Time: A Combat Surgeon in World War II* (Boston: Little, Brown and Company), p. 132.
52 M. Doubler (1994) *Closing With the Enemy: How GIs Fought the War in Europe* (Lawrence: University Press of Kansas), pp. 41–60; Cosmas and Cowdrey, *Medical Department*, pp. 364–372.

5
Day-to-Day Health

Abstract: *This chapter examines the challenges related to general, daily, combat-related health care needs apart from wounds. Cold weather, illness, disease, and combat exhaustion generated significant problems for the combat medics. They treated these and other ailments at a daily sick call for their unit even as the fighting continued. And while American troops always remained the primary focus, medics at times assisted both German combatants and civilians. Combat medics uniquely and effectively met an unexpected, broad spectrum of health care needs.*

Shilcutt, Tracy. *Infantry Combat Medics in Europe, 1944–45*, Basingstoke: Palgrave Macmillan, 2013. DOI: 10.1057/9781137347695.

Death grew to be "more or less an every day occurrence" for combat medics as the European war intensified.[1] The carnage they routinely encountered forced them to adapt quickly, treating and evacuating their wounded comrades even while struggling to maintain their own emotional balance. As they patched up horrific wounds, medical soldiers moved the maimed out of harm's way and comforted soldiers who lay beyond medicine's reach, all in the course of an ordinary combat day.

Providing prompt care and evacuation of wounded soldiers remained the fundamental task of aid men and BAS personnel, but secondary factors lent increased complexity to their jobs. Apart from wounds, front line medics treated the American fighting men for both simple and involved medical problems during daily sick call. Beyond this, as time and circumstances allowed, medics also tended to civilians' health needs and treated non-American wounded as well, but only after seeing to the needs of American troops.

Aid men and BAS personnel intermittently responded to calls from non-American combatants and civilians whom war's fortunes cast into their paths. Combat medics assisted German soldiers in two unrelated ways: they offered a safe means for surrender and they rendered first aid to the wounded. When German soldiers did surrender they often sought out the Americans who displayed the red cross and yielded confidently to them, trusting medics to hold to their non-combatant status and escort them safely to prisoner collecting points.[2] Accepting the surrender of an unarmed enemy caused little worry, but medics remained cautious when treating enemy wounded. Private Glen Mitchell's 87th Infantry Division litter squad chanced upon a wounded German soldier who was being tended by his captain. The officer surrendered to the medics, and Mitchell's team stabilized the wounded soldier. As they placed the wounded man on a stretcher for removal from the field, Mitchell feared that their retreat path might be mined. So the medics signaled the German officer to lead the way, "hoping that if there were mines he would set them off first."[3] The Americans followed, "walking on eggs," and reached the BAS unharmed.[4]

Just as combat medics remained guarded when working on German casualties, so BAS teams sometimes viewed enemy wounded with apathy, showing scant interest in preserving German life.[5] One battalion surgeon recalled his own emotional tension and unprofessional attitude stirred by enemy wounded who processed through his station. At first he maintained a posture of indifference toward the Germans, treating them no differently, other than attending to American casualties prior to German wounded. But as his war drove on, he grew increasingly antagonistic toward the enemy. Eventually he delegated care of all enemy to enlisted personnel and only reluctantly reviewed their request for consultation or approval. When his MAC officer cautioned him that he "might end up on the other side of the bench before all this stuff played out," the doctor returned to a "more reasonable track," conducting himself with greater balance.[6] While this surgeon's internal conflict may not reflect the attitude of most caregivers toward enemy wounded, his honest evaluation reveals the emotional turmoil of one medical soldier who systematically and daily confronted mangled flesh.[7]

Treating German civilians prompted less disquietude among company aid men and BAS personnel who ordinarily adopted a willing attitude toward this task. As the BAS kept pace with the combat unit, they often located their station near or within villages, and if their unit remained stationary for a time, the medical soldiers opened their doors to the townspeople. Women and children of course comprised the bulk of the population, and they approached the BAS with a "variety of requests," including dressings for minor injuries, aid for respiratory infections, occasional obstetric duties, and delousing.[8]

The mobile nature of front echelon medical work, however, meant that only rarely could medics pause to treat war wounded civilians, and then usually in hurried fashion. While combat medics dedicated some time and material resources to aiding German non-combatants and wounded, this commitment represented a quite limited component of their activities.[9] With priority unflinchingly focused on treating GIs, the most pressing concern beyond combat wounds involved maintaining the day-to-day health of the American fighting troops.

Combat did not preclude the need for daily sick call. Whether the battle raged or not, front line soldiers suffered from health-related problems ranging from general ills to more serious or chronic diseases.[10] The company aid man was the "genuine Doc to the boys," usually the first to see a sick GI.[11] Their training and provisions prepared combat medics to treat minor ailments such as headaches, common colds, and blisters, but if the sick man's condition demanded more advanced diagnosis or should supplies prove inadequate to the task, then the aid man sent the soldier back to the BAS. Since the mobility of the BAS prohibited long-term therapy, when the battalion surgeon could offer no substantial relief, the ill soldier moved further back for more definitive treatment. But just as they cared for the wounded, front echelon medical men daily served as all-purpose healers, keeping their comrades fit to fight—no small task in a combat soldier's world.[12]

Infantry troops in Europe lived in miserable conditions inviting a broad range of ailments stemming from poor hygiene. One 44th Division aid man lamented that combat forced his platoon to go about "caked with mud from head to toe" and only allowed them to change "oily black" underwear perhaps twice a month, and his assessment portrays the condition of infantrymen as a group.[13] Washing was difficult at best, and troops on the line for over 30 days straight simply "got dirty and stayed dirty."[14] Infantrymen suffered from "perpetual diarrhea" and the marriage of mortars and slit trenches left GIs looking like contestants in sack races as the cough of artillery sent them scrambling from the latrine area, yanking at their pants while in a dead run.[15]

The 44th Division medical report for November 1944 testified to the broad range of digestive system ailments within infantry units across Europe. The Medical Department listed 70 percent of the troops as "affected by a moderately severe form of gastro-enteritis of unknown etymology" during the month.[16] While this may suggest some sort of water contamination or food poisoning in the 44th Division's specific instance, medics regularly helped soldiers complaining of stomach problems, some perhaps due to combat stress. One platoon leader recalled that as a line replacement, a persistent stomach ache plagued him, so his medic "fixed up a bismuth and paregoric combo [that he] sipped from time to time to get rid of the belly ache."[17]

Medics also habitually treated hepatitis, another debilitating ailment directly related to the foul living conditions.[18] Primarily a result of fecal contamination and uncertain hygiene, this disease could also be spread through infected blood, and particularly troubled the medics.[19] Hepatitis substantially reduced the fighting strength of at least one combat unit in the spring of 1945 between the battles at Bitche and at Heilbronn. Hepatitis threatened to erode medical care for the 3d Battalion, 398th Infantry when the surgeon and his assistant contracted the disease. The two evacuated within days of one another, leaving only aid station enlisted men to carry on with primary care and evacuation directives. The absence of the medical officers forced the regimental surgeon to step in and respond to the most critical needs.[20] Other infantry divisions also suffered with hepatitis in "sufficient number to cause some commanders considerable alarm," and battalion aid stations across Europe increased sanitation inspections in forward companies in an effort to gain some control of this problem.[21]

Along with hygiene-related maladies, medics found that cold weather also multiplied infantry non-combat casualties.[22] When battle lines remained static, front line soldiers rarely left their foxholes once they dug in. These in-ground redoubts invariably reached Europe's generally shallow water table, and even a carefully trenched pit husbanded rain, sleet and snow, turning foxholes into filthy cavities of mud and ice.[23] Water-soaked blankets offered little buffer against plummeting temperatures and froze to the sides of the entrenchment if the soldier remained still for any length of time. Even the "condom-like" sleeping bags which the army belatedly sent to forward fighting elements little improved the soldiers' lots.[24] The miserable conditions and the army's utterly inept planning meant that during combat GIs never dried out or warmed up and consequently suffered endemically from respiratory illnesses, viruses, and a host of problems related to exposure and lack of appropriate equipage.[25]

Company aid men treated and routinely returned to duty those soldiers who reported lesser complaints such as coughs and colds. Medics carried little or no medications for illnesses so they dispatched the patient to the aid station if he "looked bad or sounded bad."[26] At the BAS-enlisted medical technicians administered cursory checkups and if they diagnosed only a minor problem, then they gave the GI medication, let him warm up, and sent him back to the front. Battalion surgeons examined the more seriously ill patients, usually those who coughed up dark green phlegm or complained of chest pain or of shortness of breath. Since the BAS lacked diagnostic equipment, surgeons evacuated these more seriously ill soldiers to a facility with X-ray technology. The lack of equipment at the BAS also meant that doctors generally marked FUO (fever of unknown origin) on the EMTs of those evacuees with possible pneumonia.[27]

While the exposed living conditions contributed to the incidence of respiratory and viral diseases, medics also discovered that climactic conditions combined with ineffective winter gear to produce even more widespread and debilitating cold weather problems than the winter weather alone justified. In November 1944 one medic told his family of the difficult circumstances: "the weather is bad...the mud oozes and the sleet cuts like a knife and the snow puts a false blanket over everything."[28] Forward troops suffered profoundly because army-issued boots and overcoats proved pitifully inadequate as protection against Europe's winter extremes.[29]

Although soldiers layered whatever garments they could scavenge, such makeshift efforts offered no serious defense against the implacable elements and brutal environment. Many infantrymen soldiered for days without blankets and remained on the lines for weeks before receiving winter-weight jackets and wool gloves. One rifleman insisted that the army's best provision was the woolen cap that fit under the helmet liner—"no frozen ears."[30] But the standard issue boots proved the most inefficient protection against the cold and wet weather. In a classic instance of too-little-too-late the army belatedly addressed the problem by beginning the distribution of waterproof shoe-pacs, but by that time frostbite and trench foot had needlessly depleted combat strength and prompted profound suffering. Many soldiers who had survived fearsome German 88m barrages endured lifelong ailments stemming from the unconcern or ineffectiveness of their own high command.

Stateside training had equipped company aid men with basic knowledge of cold-related injuries. They understood that noses, ears, and fingers were the most susceptible to frostbite and that a soldier who suffered from this "thermal injury" first felt a pins-and-needles sensation, and the exposed skin quickly took on a "grayish or whitish waxy appearance."[31] But feet also suffered badly as the Army equipped them with faulty, non-insulated combat boots, leading to an ineluctable destiny of wet socks and consequent frostbite.

The soldier was required to report any foot problems to the medic, but if he did not or if the medic failed to identify the "blue glass frozen" feet, the condition quickly worsened.[32] The original tingling sensation gave way to numbness, lulling the soldier into thinking that the problem had disappeared, but the lack of feeling actually marked the death of tissue. If the GI did not seek immediate attention, then the condition moved from the damaged tissues into the blood vessels themselves. Without swift treatment a restricted blood supply led to gangrene, evidenced when the damaged flesh finally warmed up and the affected tissue and underlying area turned black. This signaled permanent damage, with amputation the only remaining medical option.[33]

Aid men who served on the Siegfried Line, such as Private Earl Lovelace of the 2d Division, regularly directed platoon members to check their feet for frostbite. In the evenings Doc Lovelace moved from foxhole to foxhole admonishing each soldier to remove his boots, massage his feet, and change into dry socks.[34] If medics identified frostbite in its early stages they evacuated the soldier to the rear for a few days after which the lineman returned to the front.

Of course medics found that they enjoyed no special immunity from this ailment.[35] Shortly after Christmas 1944, one 1st Division aid man discovered that his toes had turned white and evacuated himself for a 24-hour period. Upon his arrival at battalion headquarters, he confiscated a pair of German boots for his return to the line. His feet stayed warm and dry in the enemy's effective footwear, footwear his own army had neglected to provide.[36]

Even in milder weather, soldiers who could not keep their feet dry developed a slightly different but equally painful and incapacitating circulatory disorder, trench foot.[37] If front line medics knew about this injury, it was probably recalled as a vague training memory from a casual comment or field instruction booklet. While trench foot was a major problem in World War I, the Army simply ignored its own tragic experience as the Medical Department inexplicably provided little detail on treatment apart from a brief reference, although it devoted considerable training manual space to frostbite. Labeling the problem "trench foot, shelter foot, and immersion foot," the Medical Department identified the conditions as "practically the same [and] very much like mild frostbite." The manual merely recommended dry feet and good circulation as a preventive to this disorder, a foolish suggestion for ineffectively shod foot-soldiers long on the line.[38]

Trench foot clearly proved a more substantial problem than anticipated, and it presented subtly distinct symptoms from frostbite. The skin on the feet turned pale and mottled, and the extremities seemed to lose pulse. Initially the soldier felt little pain, but if the feet remained wet the condition worsened as tissues swelled, turned red, and now-excruciating pain shot through the feet and up the legs. Trench foot incapacitated significant numbers of infantry soldiers. One lineman recalled that the disorder hit his battalion so hard that the intense, crippling pain forced his men to kneel on guard duty.[39] If the injury advanced, pain and accompanying edema made it unlikely that the affected soldier could return to front line duty; there also existed the very real possibility of amputation of toes or feet.[40]

The Army's woefully inept lack of equipage meant that the drain on manpower due to trench foot far exceeded expectations. At times, line companies' cold-related foot injury casualties approached the number of those evacuated due to combat wounds. One replacement officer stayed on line for only eight days before the crippling ailment rendered him useless for combat.[41] The "severe deterioration" of some American units was also acutely evident in the Hürtgen Forest, where the frigid weather kept the 28th Division perpetually wet and cold.[42] The 4th Division also struggled to maintain a full complement on the line in the baleful forest as one battalion evacuated 28 riflemen with trench foot within a 16-day period.[43] Even when the weather moderated, melting snow and continuing rainfall raised ground water levels and kept trench foot incidents at high numbers.[44]

First echelon medics tried several ways of battling this malady. One medical unit had enough success with a waterproofing technique they developed that they recommended it for all infantry soldiers. They had their line soldiers wear cellophane bags taken from 155mm ammunition boxes sandwiched between two pair of thin socks.[45] Other divisions tried rotating the men into warm, dry areas during static periods, but war afforded such respites so rarely that by the time the men reached the rear the tissue damage was often irreversible.[46] One 30th Division medic recommended that his platoon members use newspapers as insulation, while 398th Battalion surgeon Brown McDonald found that for his unit, "short stays in the warm kitchen area" allowed some to return to duty.[47] Company medics ran the same risks as the linemen, but some replaced their water-soaked boots with dryer ones belonging to casualties who did not "need them any longer."[48]

Throughout the divisions, riflemen commonly wore one pair of socks while keeping a second pair relatively dry by carrying them in their arm pits or inside their shirts. In Italy 88th Division, Battalion surgeon Neel Price managed to take extra pairs of socks into battle. He directed the GIs under his care to pin the extra pair inside their shirts next to their skin so that they would have a dry pair. Although they quickly exhausted the original replacement socks, Price's BAS managed to provide clean socks almost daily during the winter of 1944–45, and by his assessment, his battalion "virtually eliminated" trench foot with this precaution.

The Army command, which had so badly served its combat elements with grossly inadequate equipment and training, compounded the problem in one misguided attempt to solve it. The Army distributed rubber-soled shoe-pacs to the troops, but some linemen found that this effort only complicated existing foot problems by initiating an associated malady known as shoe-pac foot.[49] The new shoe-pac boots had waterproof soles and kept water from soaking in, but the shoe-pacs still failed to keep their feet dry. The rubber soles caused feet to sweat and the faulty construction of the boot allowed water to "just come in the top of them [making] it worse."[50]

Not until January 1945 did coherent orders to combat trench foot reach the front medical units. Trench foot control teams met with BAS personnel and aid men and laid out a preventative course of action which corresponded with the more successful attempts already underway: rotation off the line, frequent foot inspections, and spare socks. While medics helped educate their units on the importance of keeping feet dry, to stress the point some company commanders took charge of foot inspections. One 78th Division platoon leader visited every foxhole after dark and ordered his men to remove their shoes and socks. He then "squeezed the meat on the ball of the foot...to see that it was solid. If they had trench foot the meat was beginning to deteriorate and out they would go."[51]

The most severe cases required aid men to cut off shoe-pacs and combat boots when soldiers could not even remove the footwear for inspection; almost crippled by this point, these men required litter evacuations. The pain and inflammation suffered made it impossible for the soldiers to again wear boots on feet swelled like balloons. If the man recovered enough from the initial injury that the Medical Department released him as fit for duty, the damaged tissue was so susceptible to re-injury that combat responsibilities were often no longer an option. Despite the directives and the eventual introduction of the ineffective shoe-pacs, foot injuries continued to deplete the fighting force. Trench foot losses may have run as high as ten to fifteen percent of unit strength at the peak of the crisis, but inconsistencies in reportage make reliable estimates difficult.[52]

Since the Medical Department failed to standardize the procedure for characterizing trench foot losses, reports varied from one unit to the next. Although most reports recorded the case count within the "injuries" categories, medics disagreed on whether to consider the sufferer as a combat or a non-combat casualty. Units that identified trench foot as a category separate from injuries included separate notations for soldiers who evacuated with immersion foot, frostbite, or a "cold injury-ground type."[53]

Despite inconsistencies in labeling the injury, by late spring 1945, aid men began seeing fewer cases of trench foot, due probably to a combination of factors. The front line troops received a more abundant, if still inconsistent, supply of socks, medical units attempted to comply with army directives for prevention, and most importantly, the weather began to moderate. But by this point, the cruelly neglectful outfitting of the soldier's feet had caused an unanticipated and costly drain on the fighting units.[54]

If medics were under-prepared to respond to cold-induced injuries and diseases, they were even less equipped to deal with neuro-psychiatric casualties, generically referred to as combat exhaustion. This multi-faceted disorder manifested itself during all seasons, among both new and veteran troops, and across all ranks.[55] One medic called it the "saddest thing" that he dealt with, noting that some of the soldiers suffering with combat exhaustion "were previously heroic."[56] In common with the inconsistent labeling of trench foot, medical reports varied from division to division both in terminology and in understanding concerning combat exhaustion. Categories used by the different units to report the condition included exhaustion (physical/mental), neuro-psycho casualties, combat exhaustion, fatigue, psychoneurosis, and battle exhaustion.

Adding to the inconsistency of the reports, some medical units distinguished between mental exhaustion and physical exhaustion, often placing the former under the non-battle injury category and the latter within the grouping for diseases or even wounded in action (WIA).[57] Battalion surgeon Frank Ellis proposed that in the "individual immersion in the cauldron of all out war" every soldier initially possessed some level of tolerance, but once exceeded, "it initiate[d] this syndrome."[58] Another battalion surgeon simply noted that "combat fatigue was inevitable—everyone had his breaking point."[59]

For some, combat stirred a growing sense of fear and helplessness that finally prompted their evacuation because of resultant psychological problems. One front line veteran assessed the combat soldier's mental condition in commenting that "going into battle we [were] all like zombies, scared as hell, not knowing what [lay] ahead."[60] Even before the battle began the smell of death, of "rotting bodies," foreshadowed the horror to come; each man entered combat with his personal demons that often increased anxiety levels.[61] Once engaged, the noise and confusion of the fight, especially the unrelenting artillery fire, kept men in a "state of jitters."[62] Both enlisted men and officers fell victim to combat exhaustion; one company aid man reported that the shelling so terrified one of his company officers that the man would not leave the safety of his foxhole, "not even to urinate."⁶³ The officer's alarm led him to use empty whiskey bottles as urinals, which he then handed to an enlisted man to dump. The medic then stepped in, ordering the officer to the rear for a rest as a "battle fatigue" casualty.⁶⁴

For other line soldiers the "uncertainty of everything," particularly the anticipation of being wounded, fostered mental or physical collapse.⁶⁵ An assault in the Hürtgen Forest that killed 24 4th Division soldiers, which shocked the unit so profoundly that it left them on edge for several days. Shortly after the battle an explosion near a battalion ammunition dump sent a private running toward the company CP. The GI "dove head first" into the foxhole with his captain, screaming all the while that one of his legs had "been shot off and the other wounded."⁶⁶ The captain and his companions looked the young man over, all the while explaining to him that he had been "running awfully fast" to have lost a leg.⁶⁷ But the frightened soldier continued to insist that his superiors check him over. They found no wounds, but only blood sprayed from other casualties.⁶⁸

Physical deprivations also compounded feelings of helplessness. One platoon leader suggested that "even if you were convinced you were never going to be hit," exhaustion could still overwhelm the soldiers.⁶⁹ The discomfort of the foxholes, the bad weather, and night shellings allowed soldiers little rest, and many units not uncommonly got only three or four hours sleep per night. Soldiers suffering combat fatigue might be found "wandering aimlessly," with pupils dilated, unblinking, and a blank expression.⁷⁰ Although medics sometimes sent sleep-deprivation casualties to neuropsychiatry hospitals well behind the line where doctors induced sleep through drugs, most front line medical units tried to keep the combat exhausted men as close to the front as feasible.⁷¹

Battalion surgeon Brown McDonald evacuated mild cases to the kitchens where soldiers could sleep with the aid of sedatives. Other units designated rest camps, "havens of refuge," close enough to the front that the soldier could still hear the noises associated with combat.⁷² Some officers felt that moving the traumatized soldier too far from the line might "ruin them," so they ordered medics to keep the milder cases near the line as the Army allowed infantry officers full discretion in handling mental illness, an unthinkable procedure in reference to physical ailment. Although the term "rest camp" might imply an organized recovery center, the facilities were little more than dugouts or bombed-out shelters located less than a mile from the fighting. These often provided at the least hot meals, a morale boosting substitute for the C and K rations that served as the foot soldiers' main meals.[73]

Fatalism sometimes supplanted fear in combat troops as GIs watched shrapnel tear through the bodies of their comrades while they remained unharmed. As infantry soldiers settled into the endlessly violent routine of war they could become "robotic in mental acuity," convinced that when it was their time, they would be killed or wounded.[74]

Some men could find no mechanism to numb themselves against extreme fear, so they devised their own method of escape through self-inflicted wounds (SIW). Medics dealt with SIW on a case-by-case basis, since SIW proved difficult to ascertain. One battalion surgeon noticed that the increase of SIW signaled that a new battle was about to start while another connected SIW to the more intense battles. Regardless of when inflicted, most SIW were to the foot, so combat medics learned to discern whether the man was left handed or right. Inevitably the opposite foot had been shot. At least one aid man living with the men of his unit in the midst of the horror labeled SIW victims as WIA because he felt that "anyone who was that distressed and terrified was no use to us."[75] Others may well have followed this practice and kept their own counsel.

While a combination of fatalism, fear, and physical exhaustion contributed to a diagnosis of combat exhaustion, emotions such as remorse must also be considered as causative of this ill-defined neuro-psychiatric disorder. Battalion surgeon Frank Ellis suggested that for those who did kill enemy soldiers, "feelings of guilt for having killed other humans might have aggravated the situation."[76] Medics witnessed a wide range of responses among the linemen who took enemy lives. One company aid man noticed that "a pink-faced kid cried like a baby" after he killed a young German.[77] Another medic encountered a GI who developed a more brutal coping mechanism—the soldier collected ears from the Germans he had killed.[78] Aid men themselves, surgeon Ellis speculated, bore some immunity to combat exhaustion because they did not carry the responsibility of taking life; their purpose was to save, not destroy. Yet aid men did fall to combat exhaustion because even for the healers "each battle or dangerous situation took its toll, until you finally snapped."[79]

On occasion soldiers feigned combat exhaustion, but physicians dealing with clearly genuine cases found a comprehensive diagnosis almost impossible. Whether the source of the breakdown sprang from fatalism, fear, physical deprivations, guilt, or a combination of the four, battalion surgeons found "no reliable way to recognize its imminent appearance," so they charged the company aid men with identifying the signs of distress that called for immediate attention.[80] Medics must now watch closely for irritability, uncontrollable shaking, non-responsive behavior, or an inability to concentrate and complete tasks. More extreme breaks with reality appeared in men who lost their sight or speech with no discernable physiological cause.[81]

The Army now tasked combat medics, who had received at best inadequate instruction in treating war-torn bodies, with diagnosing the extraordinarily subtle manifestations of psychic disturbances—still puzzling to the most practiced psychiatrist. Aid men tried to talk some of the men through their fears but immediately evacuated those they judged the more serious cases to the BAS.[82] Battalion surgeons looked for "unexpressed but obvious behavioral evidence of flightiness and apprehension as important features" in deciding whom to evacuate and whom to send back to the line.[83] For some, a short break restored physical and emotional strength and these soldiers returned swiftly to duty, while the more severely traumatized received specific and extended treatment not possible near front lines.[84] Such para-psychological tasks typify the bewilderingly complex duty laid on the beleaguered front line medics.

In the ordinary work of their extraordinary days infantry combat medics aided Germans with various levels of enthusiasm, although first devoting themselves to their troops' health. Daily sick call gave aid men the opportunity to respond to lesser complaints while referring more serious ailments to the BAS. Since weather-related injuries, diseases, and a variety of hygienic problems could reduce substantially the fighting force, combat medics aggressively pursued ways of coping with these. Beyond this, extreme physical and mental fatigue drained manpower further, as unanticipated combat exhaustion casualties mounted. Although medics worked to intervene before men broke completely, the realities of combat proved intensely personal for each soldier, affecting each differently. The exigencies of war and an unthinking or inadequately concerned command system layered all these disparate tasks on the combat medic whose fundamental duty remained with his wounded comrades.

The Army's shortcomings in educating combat medics concerning non-combat injuries and illnesses together with its failure to supply appropriate cold-weather gear exposes the woeful lack of preparation afforded medics and soldiers alike. And while medical soldiers' mission was to care for wounded GIs, the combat environment and the army's obtuseness forced on them a multiplicity of secondary tasks absent from the training regime. Medics adapted as always as best they could, responding to peripheral needs as occasion and time arose, but always concentrating on the most important matter at hand. The extraordinarily complex nature of their tasks, unsupported by any realistic training, forced combat medics to spend themselves in meeting the needs of their units. The emotional toll on medics proved considerable.

Notes

1. Robert Reed to dad, 28 March 1945, in Reed's possession.
2. Questionnaires, Frank R. Ellis (4 June 2001), Serge Manni (25 September 2001), and Walter Murchison (8 May 2001); C. B. Currey (1984) *Follow Me and Die: The Destruction of an American Division in World War II* (New York: Stein and Day), p. 92; E. Engle (1967) *Medic: America's Medical Soldiers, Sailors, and Airmen in Peace and War* (New York: John Day Co.), p. 86; H. Kemp (1990) *The Regiment: Let the Citizens Bear Arms!* (Austin: Nortex), p. 86; A. N. Towne (2000) *Doctor Danger Forward: A World War II Memoir of a Combat Medical Aid Man, First Infantry Division*. (Jefferson, NC: McFarland and Co., Inc.), p. 118; "General Orders Number 4, 12 January 1945, After Action Reports, 110th Medical Battalion, 35th Division," Box 9790, RG 407, NARA.
3. Questionnaire, Glen J. Mitchell (22 August 2001).
4. Questionnaire, Mitchell.
5. Questionnaires, Carl R. Aschoff (14 June 2001), Stephen J. Barnett (28 June 2001), Roy Barratt (13 June 2001), William Braunhardt (30 May 2001), Ellis, Wilbur Heinold (2 April 2001), Richard L. Lease (18 May 2001), Earl Lovelace (31 August 2001), Russell Wade Redfern (1 June 2001), and Paul Winson (31 May 2001); Currey, *Follow Me*, p. 92; "44th Infantry Division, Public Relations Section, by Sergeant Bill O'Hollaren, After Action Report-Office of the Public Relations Section, 44th Division," Box 10804, RG 407, NARA.
6. Questionnaire, Ellis.
7. In his combat memoir, Allen Towne, a Medical Battalion litter bearer on detached service to a BAS, notes that a similar situation occurred in the station in which he worked. The Commanding Officer, a medical doctor, refused to treat a German soldier, blaming the enemy soldier for an American's death. Towne, *Doctor Danger*, p. 158.
8. Questionnaires, "variety" Ellis, and Barratt, Heinold, Lease, and Brown McDonald, Jr. (22 June 2001).
9. Questionnaires, Ellis, Heinold, McDonald, and Frank Miller (2 April 2001); J. Greenberg and H. C. McKeever, eds (1995) *Letters From a World War II GI* (New York: Franklin Watts), p. 83.

10 Venereal diseases (VD) generally did not pose severe health problems for the most forward elements during combat. Paul Fussell observed that "the front was the one wartime place that was sexless." Combat medics who responded to the questionnaires used in this study supported this contention, indicating that they rarely, if ever, dealt with VD on the front. Former Assistant Battalion Surgeon Carl Aschoff suggested that there were "no opportunities during combat, [but] after the war ended it was a whole different story." Yet, during a lull in the fighting, or after a break from the front, soldiers did contract sexually transmitted diseases. There is evidence that on occasion, aid men received offers of money to provide sulfa drugs for VD. One battalion surgeon noted that gonorrhea cases increased when his battalion was stationary for several days. Upon diagnosing the men, he gave them penicillin and sent them back to their units. After informing Division of the cases, an order came back "not to report them, so [he] changed the diagnosis to nasopharyngitis, acute to conform to the directive." P. Fussell (1989) *Wartime: Understanding and Behavior in the Second World War* (New York: Oxford University Press), p. 108; Questionnaires, "no opportunities" Aschoff and Barratt, and Andrew Ciampa (7 December 2001); "not to report" Frank Miller to author, 25 April 2001; J. L. Ellis (1980) *The Sharp End: The Fighting Man in World War II* (New York: Charles Scribner's Sons), pp. 182–186.

11 Questionnaires, "genuine" Ellis, and Heinold; Ellis, *Sharp End*, pp. 182–186.

12 Questionnaires, G. Allen (25 March 2001), Aschoff, Barnett, Barratt, Ellis, Thomas Hoke (18 June 2001), Carroll E. Pomplin (n.d.), and Redfern; Greenberg and McKeever, *Letters*, pp. 89–92; J. C. McManus (1998) *The Deadly Brotherhood: The American Combat Soldier in World War II* (Navato, CA: Presidio Press), p. 154; "Historical Records and History of Organization—Month of September, 30 September 1944, 1st Medical Battalion, 1st Division, European Campaign," Box 5966, RG 407, NARA; "Action Against Enemy, Reports After Action, 2 June 1945, 2d Medical Battalion, 2d Division," Box 6091, RG 407, NARA.

13 W. S. Tsuchida (1947) *Wear it Proudly: Letters* (Berkeley: University of California Press), p. 14.

14 Tsuchida, *Wear It*, p. 102; Greenberg and McKeever *Letters*, p. 92; Questionnaires, G. Allen, Walter Biggins (29 June 2001), and Redfern; R. B. Bradley (1970) *Aid Man!* (New York: Robert Bradley), p. 55; R. Holmes (1985) *Acts of War: The Behavior of Men in Battle* (New York: The Free Press), p. 215.

15 Questionnaire, "perpetual" Biggins; G. P. Arrington (1959) *Infantryman at the Front* (New York: Vantage), p. 75. Tsuchida, *Wear It*, pp. 51–52.

16 "Resume of Sanitary Report for November, 1 December 1944, Medical Department Problems and Activity, 44th Division," Box 10804, RG 407, NARA; G. A. Cosmas and A. E. Cowdrey (1992) *The Medical Department:*

Medical Service in the European Theater of Operations (Washington, DC: Center of Military History), p. 543.
17 Questionnaire, Jacob E. Way (19 April 2001).
18 "Medical Department Problems and Activity for the Month of March, 8 April 1945, 44th Division," Box 10804, RG 407, NARA; Questionnaires, Aschoff, Biggins, and McDonald; Cosmas and Cowdrey, *Medical Department*, p. 543; "Unit History, 63rd Division," RG 94, NARA.
19 C. B. Clayman ed. (1989) *The American Medical Association Encyclopedia of Medicine* (New York: Random House), pp. 532–533.
20 Questionnaires, Aschoff and McDonald.
21 "Medical Department Problems, 8 April 1945, 44th Division, Office of Surgeon, January to June, 1945," Box 10804, RG 407, NARA; "2d Battalion Aid Station Log, 69th Division, 1945," Box 11323, RG 407, NARA; "sufficient" "Medical Department Problems and Activity for the Month of March, 8 April 1945, 44th Division," NARA.
22 L. Atwell (1958) *Private* (New York: Simon and Schuster), p. 82; Ellis, *The Sharp End*, pp. 182–186; H. P. Leinbaugh and J. D. Campbell (1985) *The Men of Company K: The Autobiography of a World War II Rifle Company* (New York: William Morrow Company), p. 197; "Resume of Sanitary Report for November, 1 December 1944, 44th Division," RG 407, NARA; Cosmas and Cowdrey, *Medical Department*, pp. 488–543.
23 Questionnaires, Aschoff and Angelo Zanin (n.d); Tsuchida, *Wear It*, pp. 9, 103; Leinbaugh and Campbell, *Men of Company K*, p. 54; Interview, "Hürtgen Forest Replacements and Non Battle Causalities with Frye, et.al."; Interview, "Report Hürtgen Forest Battle, 4th Division, First US Army, 7 November to 3 December 1944, by Captain K. W. Hechler, with the 1st Battalion, 22d Infantry, Division Narrative by Lieutenant Colonel William T. Gayle, 4th Division Information and Historical Service," CI 34, Box 24021, RG 407, NARA; "Resume of Sanitary Report for November, 1 December 1944, 44th Division," RG 407, NARA.
24 Questionnaire, Way.
25 Ibid.; K. E. Bonn (1994) *When the Odds Were Even: The Vosges Mountains Campaign, October 1944* (Navato, Ca: Presidio Press), pp. 27, 118; M. Doubler (1994) *Closing With the Enemy: How GIs Fought the War in Europe* (Lawrence: University Press of Kansas), p. 240; "Resume of Sanitary Report for November, 1 December 1944, 44th Division," NARA; G. Astor (2000) *The Bloody Forest: Battle for Hürtgen, September 1944–January 1945* (Novato, Ca: Presidio Press), p. 145.
26 Questionnaires, "looked" Lovelace, and David E. Fought (7 April 2001); Pomplin, Redfern, and Robert R. Reed II (28 April 2001).
27 Questionnaires, Aschoff, Ellis, McDonald, Miller, and Neel Price (June 2005).
28 Tsuchida, *Wear It*, p. 29.

29 Questionnaires, Aschoff and Ellis; Doubler, *Closing With the Enemy*, pp. 240–242; Bonn, *When the Odds Were Even*, p. 26.
30 Questionnaires, "no frozen ears" Way, and Aschoff and Ellis; Interviews, "Report Hürtgen Forest Battle, 4th Division by Hechler and Gayle," RG 407, NARA; "Resume of Sanitary Report for November, 1 December 1944, 44th Division," RG 407, NARA; D. Fought, "Memories of a 1st Division Medic in World War II," *Bridgehead Sentinel* Society of the First Infantry Division (Summer 2000).
31 Questionnaire, "thermal" Ellis; "grayish" Fought, "Memories"; Clayman, *Encyclopedia*, p. 469.
32 FM 21-11 (7 April 1943) "Basic Field Manual: First Aid for Soldiers" (Washington, DC: Government Printing Office), pp. 53–57; Questionnaires, James Hanson (31 March 2001), Lovelace, Way, and Winson; "blue-glass" B. Phipps (1987) *The Other Side of Time: A Combat Surgeon in World War II* (Boston: Little, Brown and Company), p. 6; Fought, "Memories."
33 Leinbaugh and Campbell, *Men of Company K*, pp. 166–167; Cosmas and Cowdrey, *Medical Department*, p. 489; Clayman, *Encyclopedia*, p. 469.
34 Questionnaire, Lovelace.
35 Fought, "Memories"; "Action Against Enemy, Report After, 4 February 1945, 35th Division," Box 9623, RG 407, NARA; "Medical Department Problems and Activity for the Month of January, 1 February 1945, 44th Division," RG 407, NARA.
36 Fought, "Memories"; Questionnaire, Fought.
37 Questionnaires, "circulatory" McDonald, and Aschoff, Ellis, Lease, and Mitchell; "Resume of Sanitary Report for November, 1 December 1944, 44th Division," RG 407, NARA.
38 FM 21-11, pp. 53–57; Ellis, *Sharp End*, p. 186.
39 Currey, *Follow Me*, p. 246; H. T. Hyman (1965) *Handbook of Differential Diagnosis* (London: Pitman Medical Publishing, Co. Ltd), p. 286; R. Cotran, ed. (1994) *Robbins Pathological Basis of Disease* (Philadelphia: WB Saunders Company), pp. 400–401; Ellis, *Sharp End*, p. 186; "Periodic Report, 8th Medical Battalion, 1 January to 1 June 1945," pp. 3–34, RG 407, NARA; Interview, "Lieutenant Colonel David O. Byars, 2d Battalion, 328th Infantry, 26th Division, by Lieutenant Stockton, 55th General Hospital," HI, Box 24240, RG 407, NARA; Tsuchida, *Wear It*, p. 9; Atwell, *Private*, p. 82; Leinbaugh and Campbell, *The Men of Company K*, 54; "Hürtgen Forest Replacements and Non-Battle Casualties, 1st Battalion, 22d Infantry, 4th Division, 16 November to 3 December 1944, with Captain Jennings Frye, S-1 1st Battalion, Lieutenant George Kozmetsky, Assistant Surgeon, 1st Battalion, Technical Sergeant 3rd Grade Harry I. Fingerroth, 1st Battalion Aid Station, Vicinity Gostingen, Luxembourg, 20 December 1944, Interviews by Captain K.W. Hechler, 2nd Information and Historical Services (VIII Corps)," CI

40 34, Folder II, Box 24021, RG 407, NARA; Interview, "2d Lieutenant Donald E. Martini, Company C, 110th Infantry, 28th Division by Lieutenant Posvar, 108th General Hospital," HI, Box 2421, RG 407, NARA; Questionnaire, John Sullivan (n.d.).
40 Hyman, *Handbook*, p. 286; Cotran, *Robbins*, pp. 400–401; Ellis, *Sharp End*, p. 186.
41 "Medical Department Problems and Activity for the Month of January, 1 February 1945, 44th Division," Box 10804, RG 407, NARA; Leinbaugh and Campbell, *Men of Company K*, 54; Interview, "Byars by Stockton," RG 407, NARA; Interview, "Lieutenant Bertram Saymon, Company L, 110th Infantry, 28th Division," HI, Box 2421, RG 407, NARA; Questionnaire, Len Karlin (17 July 2001); "Resume of Sanitary Report for November, 1 December 1944, 44th Division," RG 407, NARA; Interview, "1st Lieutenant Samuel L. Evans, Company E, 104th Infantry, 26th Division, by Captain Phelps, 22d General Hospital," HI, Box 2421, RG 407, NARA.
42 Interview, "Martini by Posvar," RG 407, NARA.
43 Interviews, "Interview, 2d Lieutenant Donald E. Martini, Company C, 110th Infantry, 28th Division by Lieutenant Posvar, 108th General Hospital," HI, Box 2421, RG 407, NARA; Interview, "Hürtgen Forest Replacements and Non Battle Causalities with Frye, et.al.," RG 407, NARA.
44 Leinbaugh and Campbell, *Men of Company K*, pp. 54 and 187; Questionnaire, Lease.
45 "Battle Experiences," Number 66, February 1945, RG 407, NARA.
46 Ibid.; Leinbaugh and Campbell, *Men of Company K*, p. 54; Questionnaire, Ellis; "Battle Experiences, European Theater of Operations, United States Army, Number 15, 19 December 1944," Number 30, 5 January 1945, RG 407, NARA.
47 Questionnaires, Ciampa, and "short" McDonald.
48 Questionnaire, Karlin.
49 Questionnaire, "virtually eliminated" Price. In *The Medical Department*, Cosmas and Cowdrey note that soldiers who received the shoe-pacs during mild weather thought them superfluous and discarded them. Once the army realized the crisis proportions of the problem, they attempted to procure additional boots in December 1944. Shipping and production delays kept the boots from reaching the front lines until the spring of 1945, after the crisis had passed. Cosmas and Cowdrey pp. 492–493; "Medical Department Problems and Activity for the Month of January, 1 February 1945, 44th Division," RG 407, NARA; Questionnaires, Aschoff, Lease, Reed, and Sullivan; "Resume of Sanitary Report for November, 1 December 1944, 44th Division," RG 407, NARA.
50 Interview, "just come" "Byars by Stockton," RG 407, NARA; Questionnaire, Way.

51 Questionnaires, "squeezed" Way, and Aschoff, Bossidy, Ellis, McDonald, and Sullivan; Doubler, *Closing With the Enemy*, pp. 241–242; "Medical Department Problems and Activity for the Month of February, 3 March 1945, 44th Division," Box 10804, RG 407, NARA; "Summary of G-1 Activities for the period February 1 to 1 March 1945, 66th Division," 5 March 1945, Box 11286, RG 407, NARA; Leinbaugh and Campbell, *Men of Company K*, p. 54; "2nd BAS log, 69th Division," RG 407, NARA; Cosmas and Cowdrey, *Medical Department*, pp. 488–506.

52 Percentage taken from Cosmas and Cowdrey, *Medical Department*, p. 494; "Periodic Report, 8th Medical Battalion, 1 January to 1 June 1945," pp. 3–34, RG 407, NARA; Currey, *Follow Me*, p. 246; Leinbaugh and Campbell, *Men of Company K*, p. 167; "Resume of Sanitary Report for November, 1 December 1944, 44th Division," RG 407, NARA.

53 Cosmas and Cowdrey, *Medical Department*, pp. 494–506; Questionnaires, Aschoff, Lease, Lovelace, Sullivan, and Way; "Unit History, German Campaign, Medical Section, 45th Division," 7 April 1945, Box 10990, RG 407, NARA; "2d BAS log, 69th Division," RG 407, NARA; "Resume of Sanitary Report for November, 1 December 1944, 44th Division," RG 407, NARA; "Medical History of 69th Division, 2 February to 9 February 1945," Weekly File Number 17, Box 11323, RG 407, NARA.

54 Cosmas and Cowdrey, *Medical Department*, pp. 494–506; Questionnaires, Aschoff, Lease, Lovelace, Sullivan, and Way; "Unit History, German Campaign, Medical Section, 45th Division," 7 April 1945, Box 10990, RG 407, NARA; "2d BAS log, 69th Division," RG 407, NARA; "Resume of Sanitary Report for November, 1 December 1944, 44th Division," RG 407, NARA; "Medical History of 69th Division, 2 February to 9 February 1945," Weekly File Number 17, Box 11323, RG 407, NARA.

55 Questionnaires, Kenneth T. Delaney (n.d.), McDonald, Murchison, Walker Powe (16 August 2001), and Redfern.

56 Questionnaire, Karlin.

57 Interview, "Officers, 2d Battalion, 12th Infantry, 4th Division, 7 August 1944, Folder 31, Box 24021, RG 407, NARA; Interview, "Anderson and Piper by Fife," RG 407, NARA; Doubler, *Closing with the Enemy*, p. 242; Fussell, *Wartime*, pp. 281–282; E. D. Cooke (1946) *All But Me and Thee: Psychiatry at the Foxhole Level* (Washington, DC: Infantry Journal Press), pp. 11–18; P. S. Kindsvatter (1998) "Doughboys, GIs and Grunts: Fear, Resentment and Enthusiasm in the Combat Zone," (PhD Dissertation, Temple University), p. 496; "Action Against Enemy, Reports After Action, 2 June 1945, 2d Medical Battalion," RG 407, NARA; "Historical Records and History of Organization-Month of September, 30 September 1944, 1st Medical Battalion; Report of Activities, Company A, 9th Medical Battalion, 9th Division, 3 July 1944," Box 7545, RG 407, NARA; "Patients treated at this station, Company C, 9th

Medical Battalion, 9th Division, 4 July 1944," Box 7547, RG 407, NARA; "History of Medical and Surgical Care Given by Company D, 9th Medical Battalion, 9th Division, 1 July 1944," Box 7545, RG 407, NARA.
58 Questionnaire, Ellis.
59 Questionnaire, "combat fatigue" Powe. On the theme of inevitability, see also Questionnaires, Karlin and McDonald; P. Boesch (1962) *Road to Huertgen: Forest in Hell* (Houston: Gulf Publishing), pp. 216 and 244; J. Bourke (1999) *An Intimate History of Killing* (New York: Basic Books), pp. 234–235; Interview, "Lieutenant Colonel Kenneth R. Lindner, Commanding Officer, 3rd Battalion, Major Herman R. Rice, Jr., Executive Officer, 3d Battalion, 1st Lieutenant Cornelius R. O'Donnell, et.al., 16–24 December 1944, 3d Battalion, 12th Infantry (right flank) 4th Division, Battle of Luxembourg, Osweiler and Dickweiler, by Lieutenant Colonel William T. Gayle, Captain D. G. Dayton, and Lieutenant S. J. Tobin," 27 December 1944, Box 24022, RG 407, NARA; NARA; D. Grossman, *On Killing: The Psychological Cost of Learning to Kill in War and Society* (Boston: Little Brown, 1995), pp. 43–83, Kindsvatter, "Doughboys," pp. 499–513.
60 Questionnaire, Zanin.
61 Questionnaire, Biggins.
62 Interview, "state" "Officers," NARA; Questionnaires, Barratt, Lease, and Everett Smith (7 August 2001); Interview, "Battle of Hürtgen Forest, 16 November–3 December 1944, Battalion Aid Men 1st Battalion, 22d Infantry Regiment, 4th Division, with Lieutenant George Kozmetsky, Assistant Surgeon, 1st Battalion, Technical Sergeant, Third Grade Harry I. Fingerroth, 1st Battalion Aid Station, Technical Sergeant, Fifth Grade Joseph J. Thomas, Aid Man with Company B, Technical Sergeant, Fifth Grade Wade H. Carpenter, Aid Man with Company B, by Captain K.W. Hechler, 20 December 1944," CI, Box 24021, RG 407, NARA; Interview, "Marshall O. Miller, 53rd General Hospital, 331st Infantry, 83d Division," HI, Box 2235, RG 407, NARA; Interview, "Sergeant Joseph Sluzis, 302d Infantry, 94th Division," Box 24240, RG 407, NARA; Interview, "Anderson and Piper by Fife," RG 407, NARA; Boesch, *Road to Huertgen*, 21; D. Chernitsky ed. (1991) *The Foxholes: By the Men of the 110th Infantry, World War II* (Uniontown, PA: Dorothy Chernitsky), pp. 231–235; S. Stouffer et al. (1949) *Studies in Psychology in World War II*, Volume II (Princeton: Princeton University Press), pp. 70–88.
63 Questionnaire, Mitchell; "not even" R. L. Sanner (1995) *Combat Medic Memoirs: Personal World War II Writings and Pictures* (Clemson, SC: Rennas Productions), p. 95.
64 Sanner, *Memoirs*, p. 95.
65 Questionnaire, Smith.
66 Interview, "Captain Donald Faulkner, Company E, 22d Infantry, 4th Division, 16 December 1944," Box 24021, RG 407, NARA.

67 Interview, "Faulkner," RG 407, NARA.
68 Ibid.
69 Questionnaire, Way.
70 Questionnaire, Price.
71 Questionnaires, Smith and Redfern; Robert Reed to dad, Thursday 1945, in Reed's possession; Interviews, "Officers," RG 407, NARA; Arrington *Infantryman*, p. 60; Doubler, *Closing With the Enemy*, p. 242; W. S. Boice (1959) *A History of the Twenty-Second United States Infantry in World War II* (np), pp. 64–65; Kindsvatter, "Doughboys," p. 523.
72 Questionnaire McDonald; "havens" "Battle Experiences" Number 30, 5 January 1945, RG 407, NARA. The Army also established rest or exhaustion centers at division and army level for those considered too damaged to remain at regimental level.
73 Interview, "ruin" "Lindner, et.al. by Gayle, et.al.," RG 407, NARA; "Unit Report Number 3, 1 September to 30 September 1944, 103d Medical Battalion," RG 407, NARA; Questionnaires, Ellis, Allen L. Johnson (n.d.), Lovelace, Manni, Way, and Winson; "Battle Experiences" Number 30, 5 January 1945, RG 407, NARA.
74 Questionnaires, "robotic" Biggins, and Smith; Fussell, *Wartime*, pp. 281–282.
75 Questionnaires, "anyone" Karlin, and Barratt, T. William Bossidy (1 July 2001), Ciampa, Lovelace, McDonald, Murchison, Pomplin, Powe, Price, Richard Quint (20 November 2001), Reed, and Way; Interview, "Captain William L. Johnston, 100th Evacuation Hospital," HI, Box 24240, RG 407, NARA; Atwell, *Private*, p. 166; Bradley, *Aid Man!*, p. 55; Leinbaugh and Campbell, *Men of Company K*, p. 53.
76 Questionnaire, "feelings" Ellis; Bourke, *An Intimate History*, p. 236.
77 Questionnaire, Heinold.
78 Towne, *Doctor Danger*, p. 158.
79 Questionnaires, "each battle" Heinold, and Ellis, McDonald, and Powe. Former Battalion Surgeon Brown McDonald agrees that aid men experienced less combat exhaustion, but indicates that it was due to their identification with the line units. See Chapter 6 concerning coping mechanisms of the front-line medics.
80 Questionnaires, "no reliable" Ellis, and Bossidy.
81 Questionnaires Bossidy, Ciampa, Ellis, and Price; Towne, *Doctor Danger*, p. 150.
82 Questionnaires Bossidy, Ciampa, Ellis, and Price; Towne, *Doctor Danger*, p. 150.
83 Questionnaire, Ellis.
84 Questionnaires, Aschoff, Barratt, Ellis, Lovelace, McDonald, Miller, Murchison, Powe, and Winson; "Medical Department Problems and Activity for the Month of January, 1 February 1945, 44th Division," NARA; Bradley,

Aid Man!, p. 55; "Resume of Sanitary Report for November, 1 December 1944, 44th Division," NARA; Interview, "Lindner, et.al. by Gayle et.al.," RG 407, NARA; Boice, *A History of the 22nd*, p. 65; Bourke, *An Intimate History*, pp. 232–245; "General Orders Number 2, 24 April 1945, 103rd Medical Battalion, 28th Division," Box 8621, RG 407, NARA; "Unit Report Number 3, 1 September to 30 September 1944, 103d Medical Battalion, 28th Division, 1 October 1944," Box 8621, RG 407, NARA.

6
Company Aid Men

Abstract: *This chapter evaluates the singular relationship between the company aid man and his line unit. In contrast to the BAS soldiers, aid men labored under fire and alone. The combat units accepted the aid man as a vital component of the team, yet the aid man existed at some remove, living in a paradoxical world where conflict raged both within and without, which he dealt with in psychological solitude. The searing realities of tending combat wounded under fire translated for the aid man into a pronounced disconnect, both physically and emotionally from those around him. Remote from the BAS medics, yet bearing no weapon like his infantry fellows, the aid man labored alone, the most isolated of all combat soldiers.*

Shilcutt, Tracy. *Infantry Combat Medics in Europe, 1944–45*, Basingstoke: Palgrave Macmillan, 2013. DOI: 10.1057/9781137347695.

Doc Paul Winson and his rifle platoon moved into the Belgian village of Thirimont, knowing that the Germans had inflicted "exceptionally high" casualties. Even with this knowledge, the row of dead American soldiers propped against the walls of the hamlet's buildings startled him. Medics had treated the GIs and placed them all in sitting positions, ready for evacuation, but the bitter cold had taken its own ghastly toll before litter bearers arrived. War's harsh reality meant that the best efforts of combat medics hinged upon the system immediately behind them, and that system had failed these frozen soldiers. As a front line aid man, Winson agonized over the tragically flawed evacuation process but the horrific scene at Thirimont only hardened his determination to care for the men of his company.[1]

When he had joined the 30th Division as a replacement medic, two sergeants had dispelled Winson's obvious anxiety with a practical joke of initiation into their band. This incident set the tone for his relationship to the platoon, and "Doc Winson" came to know the men intimately, sharing every aspect of their daily lives and meeting all their medical needs. Perhaps as with most company aid men, his bond with the infantry troops superseded his official designation as a member of the battalion medical detachment, and he connected fundamentally with the foot soldiers, rarely even seeing the medical soldiers at the BAS.

Yet, the death and human detritus, which became Winson's constant companions, soon forced him to distance himself emotionally from the men of his own unit. "More than reluctant to get close to anyone," he avoided intimate situations even including "picture showing time."[2] While the platoon members affirmed Winson as an integral part of the combat team, even rewarding him with a unit promotion to a nonexistent rank, the nature of his work physically separated him from the squad for varying lengths of time. As the unit's medical specialist Doc Winson earned respect as a good soldier, but he soldiered alone.[3]

His job's autonomy sometimes put Private Winson at odds with officers from other units. On one occasion when his platoon suffered substantial losses, Winson could not get to all the wounded men, so he called a near-by field artillery observer (FAO) to help him. The FAO indicated his status as an officer and rebuffed the medic. The exasperated Winson approached the FAO, leaned close, blew breath on the man's shoulder, and pantomimed shining the missing rank insignia. The FAO still refused to assist Winson, but at least let the enlisted man's defiant insubordination pass without response. In still another episode, a lieutenant whom Winson did not know ordered the aid man to stop treating a wounded German soldier. Winson grimly told the officer to "go to hell," and continued his task.[4] Again his boldness bought no repercussions. Caring for the wounded carried certain immunity, but there was no inoculation against the task's innate loneliness.

Although each aid man's experience involved a highly personal association with his combat unit, Winson's journey exposes the inherent tensions which developed, isolating aid men from those whom they served. In common with Winson, company aid men endured the daily realities of combat with the front line soldiers. The independent nature of his work separated him both physically and emotionally from the combat troops even though the line medic identified primarily with his infantry unit. The line soldiers welcomed the company aid man as a uniquely valued team member, certainly one of their own, but the singular nature of his activities and the tactile constancy of dying and death exiled the combat field medic to a solitary realm.

The front line troops' fraternal acceptance of aid men as "part and parcel" of the team came only by an attitudinal change on the part of the combat soldiers.[5] Company aid men trained stateside with the infantry regiments, but the riflemen perceived a substantial inequality between the combat and the medical programs. In training camp medics enjoyed an easier daily regimen including exemption from guard duty and certain housekeeping tasks. Infantry trainees also believed that medical officers demanded less discipline of their men, and that the non-combatants did very little on maneuvers. Tending to others' blisters commanded no respect.

The seemingly lax approach to combat preparation prompted many infantry troops to tag the men whom they felt "had it soft" with derogatory monikers such as, "pill rollers" and "gold bricks."[6] Riflemen in training clearly did not accept aid men as "full partners."[7] But battle immediately and radically altered these disdainful appraisals of medical soldiers.

Before the baptism of fire, neither the combat troops nor the medics themselves understood the gruesome tasks which lay ahead. But following the first combat exposure, riflemen not only welcomed their aid men as comrades, but they also acclaimed them as first-class soldiers. Linemen immediately forgot any training-related trivializing or antagonism and adopted an entirely different attitude. Abandoning any derogatory nicknames, they christened the healers "Doc," and even composed verses heralding their respect for the medics, signifying that the medics just went "on and on."[8] Confident that their aid men would not desert them should they fall in battle, front troops came to lavish praise on the medical soldiers, calling them "fearless," "honeys," "selfless," and "the best."[9] Company aid man Allen Johnson rightly understood that "the riflemen worshipped" the medics.[10] The field medics earned recognition as at least equal partners in a team effort and the "infantry man absolutely respected the first aid man as he never had on maneuvers," acknowledging that medics showed all the requisite marks of soldiering.[11] One line man explained simply:

> It was hard to distinguish our platoon medic from the other men in the platoon...they looked the same, muddy boots and needing a shave. Our platoon aid man was an exceptional soldier.[12]

Line soldiers treated the aid men as "irreplaceable [and] looked out for" them in a number of ways.[13] During combat, soldiers sprayed covering fire for the medics, warned them of incoming artillery, and sheltered them until danger passed. At night linemen dug foxholes and retrieved rations for weary medics. Infantrymen offered simple signs of gratitude such as nudging their aid man to the front of the chow line, giving them extra clothes, or sharing food packages from home. On occasion, platoon officers tried to protect their medics by warning them to stay put when the firing was especially heavy.[14] But despite direct orders to lay low, medical soldiers felt compelled to help their teammates. One 35th Division medic, Gene Angelucci, baldly disregarded an order from his platoon leader to leave a badly wounded man behind; the medic made his way through a hail of fire to the injured soldier, ministered to him, and then prepared for his evacuation.[15]

In battle it seemed to the foot soldiers as if a supernatural shield surrounded some aid men, leading line soldiers at times to elevate medics to an almost super-human status.[16] Rifleman Jim Henderson thought that his aid man mysteriously "appeared to be present whenever needed," while Walter Biggins "never saw one [medic] refuse or even hesitate to get to a wounded man."[17] Some linemen actively sought the company of their medic or begged him to stay close, as if the doc's near proximity guaranteed safety.

Beyond caring for the wounded, medics also bolstered morale and lifted their fighting companions' confidence by calming anxieties of infantry replacements.[18] Doc Bradley comforted rookies who were frightened by the sounds of artillery, while other aid men took advantage of momentary lulls to reassure the new soldiers that they could indeed handle themselves in the fight ahead. Medics reminded them of their individual aid kits and urged the soldiers to use their personal supplies if wounded, promising that medical help would quickly follow.[19]

For Maurice DeLoach, assistance from his platoon medic reached far beyond physical aid. DeLoach had three brothers on the fighting line, and all four siblings had been wounded on several occasions. Each time one of the DeLoach brothers received an injury, the Army dispatched a telegram to their mother. Wounded for the fourth time, DeLoach felt that he had an ample stock of purple hearts and convinced his aid man to forego writing up the incident to prevent any more bad news home.[20]

The medical and emotional support of the company medics secured their position as valued combat team members and because the infantry so deeply respected their medics, they strenuously objected when the army declared medical soldiers ineligible for the combat infantry badge (CIB).[21] They demanded recognition for the medics and owed in large part to combat soldiers' protests, the War Department created the combat medical badge (CMB) on 1 March 1945.[22] BAS medical soldiers also qualified for the CMB and while combat troops also held them in high regard, the foot soldiers' association with the BAS medics differed from that of the aid men. While a company medic lived with his infantry platoon, BAS personnel maintained primary community with one another as fellow BAS medical soldiers. But all first echelon medics, whether on the lines or at the BAS, shared a dedication to safeguarding the combat unit.

At times the BAS officers faced their own distinctively difficult tasks in supporting their team.[23] One battalion surgeon's obligations once required him to evacuate two officers who "lost their nerve and were no longer capable of leadership or of sending their men into battle and possible destruction or death."[24] Other battalion surgeons coped with bureaucratic conflicts which at times put them at odds with their superiors.[25]

In the late spring of 1945 battalion surgeon Frank Ellis tried to return a replacement rifleman to the Maryland State Penitentiary. The soldier had volunteered for combat duty as a means to getting out of prison. After just five weeks of training he joined the war in Europe and within a week landed in Ellis's BAS. Unable to perform his duty, the man had fired less than one clip from his M1 rifle. So Ellis filled out an EMT recommending that the ex-convict return to the penitentiary as "useless for combat needs."[26] But when the soldier evacuated to the rear, Ellis's Medical Department superior overruled the doctor and sent the troubled man to a division rest camp, telling Ellis that the soldier would "be fine in a few days," thwarting the battalion surgeon's attempt to protect his battalion from potential problems.[27] BAS surgeons considered themselves as essential members of the combat team, but their medical leadership role confined their contributions "almost exclusively" to the aid station.[28]

Litter bearers who brought casualties back from the line of battle echoed the sentiments of feeling vital to the cause, but having "very little in common with the combat unit."[29] Bearers operated somewhat as independent teams in their retrieval runs, encouraged by "group pressure ... to keep going," while aid men almost always operated alone.[30] Litter bearers shaped their identity from their association with their medical squad members, but at times they also traveled with the infantry companies and developed a strong bond with the combat soldiers.[31] Too close a connection could even bring a different sort of problem. In the Hürtgen Forest an "absolutely fearless" litter bearer so enmeshed himself with the men of a rifle company that he slept, ate, and patrolled with them.[32] One cool, cloudy November day with heavy fire inflicting severe losses on his unit, the litter bearer picked up two of the wounded men, carried them on his back to the aid station, crying out "they're killing my boys, they're killing my boys."[33] He then grabbed for a gun to "overcome the feeling of helplessness" at his ability to protect his friends, but the station officer quickly evacuated him as a combat exhaustion casualty.[34]

As a rule, infantry aid men did not carry weapons, but at times some troops encouraged their medics to carry arms.[35] 2d Lieutenant Jacob Way, Jr. urged his aid man to tuck a pistol into his kit because he worried that the Germans specifically targeted medical soldiers; the medic declined and instead kept a grenade with him as a precaution.[36] Private Everett Smith who served as both a company aid man and a litter bearer found comfort in keeping a gun close by when he slept at night, but did not carry the weapon during the day.[37] During the Battle of the Bulge, Doc Paul Winson carried a .45 pistol for a few weeks after his unit "stumbled over the bodies of the soldiers murdered in ... the Malmedy Massacre."[38] The sight of the atrocity convinced him that "the SS had no intentions of taking prisoners," and although he knew he could not "hit the side of a Tiger Tank," the gun provided a degree of reassurance for a time.[39]

Some aid men volunteered to carry their squad members' weapons while moving through the countryside, and still others routinely joined patrols as riflemen. One aid man's platoon leader had him strip off his brassards before moving out on armed patrols.[40] Ordinarily, however, aid men left the fighting to the combat soldiers and relied on their protection. The front line soldiers' high regard for the aid men forged a unique relationship, but despite this, combat realities erected a wall between aid men and the members of their platoons.[41]

While company aid men remained under the direct authority of the battalion surgeon, upon assignment to a company command responsibility shifted to the line officers, who in turn allowed a great deal of freedom. Aid men welcomed the opportunity to operate independently, but that very autonomy in turn fathered a sense of abandonment and loneliness. William Braunhardt of the 100th Division joked that he "always told the sergeant that he couldn't give [me] orders," adding that as an aid man he was on his own.[42] This same lack of structure heightened Ben Burnett's impression of isolation as a company medic with the 104th Division. Burnett's training had prepared him to receive and then act on orders, but in combat his officers never issued an order or even suggested what the aid man should do. For Burnett, this self-accountability imposed a growing remoteness and a tacit understanding that his superiors did not "feel responsible" for his actions.[43] So he loyally answered every call for help, regardless of enemy fire or personal danger because he thought it his duty. Burnett enjoyed the full respect of his platoon, but his position made him captive to a disengaged existence.[44] Burnett's sober view of this solitary way of life, reflects that although he lived and worked among 40 men, he was alone. Paradoxically, some aid men relished this as independence rather than isolation; uncomfortable with taking orders, such soldiers found satisfaction acting as their "own boss almost wholly."[45]

The unique stresses inherent in their tasks further worsened the dissociation of the company medics. Aid men committed absolutely to the care of their platoon, remaining on the battlefield long after riflemen retired to their foxholes, sometimes pushing themselves beyond the breaking point. One infantry chaplain watched an aid man who landed at Normandy work for days without rest as his unit pushed into the hedgerow country. Physical and mental exhaustion finally overwhelmed the medic as he held and ministered to a young rifleman whom the chaplain termed a tragedy of war. The soldier had suffered a mortal wound but somehow clung to life, so the medic would not leave the bloodied field; his commander finally ordered him evacuated him as a combat exhaustion casualty.[46] Doc Bradley laid bare the essence of the aid man's "complete loneliness and abandonment" in his combat memoir: "The first few times you are left by yourself...are times which are profoundly shaking to your nerve and self control."[47]

Widening the gulf between the aid man and his platoon fellows, a medic's own feelings of self-doubt, inadequacy, and fear compelled him toward an even more profound emotional withdrawal. Although aid men understood that that they must make immediate decisions dictated by circumstances, they were burdened by the reality that at times "it was impossible to render aid."[48] An understanding of the limits of their own skills troubled other medics and helped isolate them from their platoon; they feared that their riflemen would lose faith in them if the foot soldiers but knew the inadequacies of medics' training.[49] Doc Carroll Pomplin expressed his feelings of helplessness and their subsequent impact:

> I will never forget the one I lost—shrapnel had cut [the] jugular vein, NO way of stopping [the] blood—He died in my arms—It hurt for a long time because his friend expected me to help and I couldn't.[50]

For other field medics, inner conflict surfaced in their efforts to control fear. One admitted that when he heard "that word medic coming from an untenable source, [his] blood [ran] cold" because he knew he had an obligation he would fulfill.[51] He saw himself as an integral component of the team, but the terror associated with combat momentarily checked his responses. Doc Frank Irgang spoke for many aid men in acknowledging the constancy of fear as he "tried to tend to required duties...tried to stay alive."[52] Private William Bossidy particularly dreaded artillery each time he worked, but decided that since he survived his first few days, that he "might make it to the end."[53] In an extreme case, panic so consumed one medic that he begged his platoon sergeant to shoot him in his leg.[54]

Not only did physical and self-imposed factors serve as isolating issues, but at times the infantrymen themselves might segregate medics emotionally from the core unit. Even calling medical soldiers "Doc" created an inherent, if unintentional, attempt to maintain distance. Platoon members addressed their officers as "Sir" or by rank and called privates by name; "the guy who carrie[d] the sulfa and bandages" lost name and rank, and was known usually as Doc, or even more anonymously as Medic.[55]

Perhaps by recognizing the aid men by name, the infantry soldiers chanced a too personal relationship which death could quickly end. One rifleman acknowledged "we called them all Doc... I don't remember the name of any of those that served with us, can't even picture their faces."[56] He further suggested that the aid men kept apart from the group for "their own mental well being [so] we did not socialize with them."[57] It may be that riflemen felt a degree of guilt for sheltering in the relative safety of their foxholes even as medics plunged directly into the line of fire; by distancing themselves from their medics, line soldiers perhaps assuaged their own self-imposed guilt. Whether consciously or unconsciously, and even though the platoon highly valued them as essential to the team, the distance widened between the rifleman and the medic.

The unrelenting involvement with the maimed and dying forced further detachment from the platoon unit as aid men raised emotional barriers against their stresses. Some medical soldiers embraced fatalism as their defense mechanism against the loneliness and solitude of their job, accepting that they "would either be killed or wounded," or anticipating that "the other guy was going to get it."[58] Company aid men rationalized that because they were at the end of the platoon, "bringing up the rear," they were not in danger.[59] Commonly aid men simply lost themselves in their work.[60] Doc Carroll Pomplin succinctly commented that he "had a job to do" and that by concentrating on the job he maintained some distance from the unit.[61]

Other medics interwove a spiritual component with other coping devices to deal with the profound disconnect born of hours working alone on the battlefield. For many, God served as the bridge between the community of the living and that of the dying. Many medics developed a greater understanding of God or simply released themselves to their creator, acknowledging that He was in control.[62] In letters home, 44th Division medic Frank Tsuchida confessed his abstracted emotions over his experiences. He described the war as a "mess" which had thrown his mind into "one confused conglomeration of incidents."[63] The medic implored his family's prayers for his unit.[64] Other medical soldiers held fast to prayer as their avenue of hope in an otherwise hopeless existence.[65] Aid man Robert Reed wrote of his renewed faith to his mother, telling her that "your middle size son went to church [where] the pews were our helmets... It relieves you to be able to tell your troubles to someone."[66]

Although some aid men did come to terms with their anxieties, growing close to and depending upon their platoon members, others steadfastly refused to assimilate into the group. Doc Allen Johnson landed with his company at Normandy, and after caring for them during their month-long combat initiation, he counted just one of the original band still at his side. This combat experience suffocated any desire he might have nourished to bond with the riflemen. The group's revolving membership meant that from that point forward Johnson "never got to know any rifleman very well."[67] He expressed a common medic sentiment in saying that he knew the team accepted him as a vital member, and that he would have done anything for the group, but he could view himself only as auxiliary.[68] Other medics suggested that although their soldiers would have done anything for them, they could not afford to get too close to any one soldier because of the pain associated with repeatedly losing friends.[69]

On occasion, this unavoidable remoteness and their job's unrelieved carnage broke company aid men physically and emotionally, yet others suffered from a paradoxical dichotomy of isolation imposed upon intimacy. While some evacuated out for a few days' rest and then returned to their units, Doc Wilbur Heinold's experience reflects the extreme psychological penalty suffered by some aid men whose intimacy with their unit prohibited them from reconciling the cost of combat.[70] In March 1945, Heinold's severe headache so consumed him that he could barely function. Evacuated with combat exhaustion, he arrived at a hospital "full of psycho cases."[71] From the time that his platoon went on line in October 1944, he had been physically apart from them just three days. His combat unit enfolded him as a vital cog in the team, and he came to wield unorthodox influence within the platoon. Superiors recognized his authority regarding the sick and wounded men, and despite his lower rank, officers complied with his demands while treating the wounded.

Although judging his training as inadequate, Doc Heinold applied common sense and field experience in carrying out his duties and also recognized that his presence kept morale at a tolerable level. As an original member of the unit he knew each soldier intimately; he understood that had he been a replacement medic "it would have been a lot easier to deal with a stranger's death" than with the personal loss he suffered when his friends died.[72] Care for the wounded consumed him, and the blood of his comrades remained on his hands and clothes for weeks at a time, an indelible reminder of his losses. Although he nurtured an intense and personal relationship with God, Heinold's six months of unrelieved care for his bloodied comrades finally took its toll, and he fought a blinding headache in a hospital behind the lines.

Standard psychological procedure of the time required that Heinold relive the terrors of combat in order to purge his mind of the trauma. So, as he lay strapped to a bed, a therapist injected sodium pentothal and played an audio recording of a combat river crossing. Heinold listened in horror and shock as a narrator shouted pretended orders, his mind struggling against the severity of his memories. He fought viciously against the restraints that held him, bruising his body to match his already suffering mind. Unable to detach from his friends while laboring with them, his attachment now became profoundly debilitating under such questionable treatment. Heinold never rejoined his platoon.[73]

Company aid men like Doc Heinold served in the most forward combat medical positions and earned a profound respect and admiration from their infantry troops which they had never gained during training. In contrast to the medical soldiers at the BAS, company aid men operated apart from like-tasked soldiers and lived the life of a combat rifleman. Line troops welcomed the aid men as invaluable team members and as fellow soldiers, understanding their mutual dependence. Linemen protected and helped them, while the aid men built esprit de corps by tending to riflemen's health needs.

Despite the aid man's secured place in community, several factors worked to separate the aid man from his core unit. His autonomy in the command structure left the medic adrift in a highly structured organization which centered on command and response, while the solitude and loneliness associated with their work's nature disconnected them from the very combat team they served. All this drained the aid men emotionally, and as they suffered continuous loss of their comrades the medics edged to peripheral positions within their small combat community. Self-imposed, internal defenses sometimes proved valuable in coping with stress, but these in turn further distances the aid men from their soldiers. Ultimately the emotional and physical detachment from the platoons cast the company aid man into a singular existence as both the most isolated medical soldier and, more profoundly, the most isolated infantry soldier.

Notes

1   Questionnaire, Paul Winson (31 May 2001).

2. Questionnaire, Winson.
3. Ibid.
4. Ibid.
5. Questionnaires, "part and parcel" John W. Rheney (10 March 2001) and Wray Richardson Thomas (13 June 2001).
6. A limited number of infantrymen who responded to the questionnaires report that they respected the medical soldiers in training, just as they did in combat. Questionnaires, "had it soft" Gerald Allen (25 March 2001) and "pill rollers" Roy Barratt (13 June 2001), Francis Irvin Berry (29 May 2001), T. William Bossidy (1 July 2001), William Braunhardt (30 May 2001), Ben Burnett (3 July 2001), Frank R. Ellis (4 June 2001), Allen L. Johnson (n.d.), Frank J. Irgang (March 2001), Richard L. Lease (18 May 2001), Victor Nash (21 May 2001), Warren C. Platt (October 2001), and Karl Stelljes (24 May 2001); "gold bricks" Burnett to author 28 April 2001, in author's possession.
7. Questionnaire, Ellis.
8. Ibid.; Questionnaires, G. Allen, Barratt, Berry, Bossidy, Braunhardt, Maurice DeLoach (20 May 2001), David E. Fought (7 April 2001), Irgang, Johnson, Lease, Serge Manni (25 September 2001), Carroll E. Pomplin (n.d.), Richard Quint (20 November 2001), Russell Wade Redfern (1 June 2001), Rheney, Stelljes, John T. Sullivan (n.d.), Thomas, and Winson; S. Stouffer et al. (1949) *Studies in Psychology in World War II*, Volume II (Princeton: Princeton University Press), p. 144; P. Boesch (1962) *Road to Huertgen: Forest in Hell* (Houston: Gulf Publishing), p. 124; Burnett to author 28 April 2001; Ben Burnett to author, 7 May 2001, in author's possession; G. P. Arrington (1959) *Infantryman at the Front* (New York: Vantage), p. 32; "on and on" Dick Patton, in "Golden Acorn News," (December 1995) 87th Infantry Division Association, James Amor, editor.
9. Questionnaires, "fearless" Biggins, "selfless" Stelljes, "the best" Kenneth T. Delaney (n.d.), and Burnett and Manni; Interview, "honeys" Interview, "Sergeant Willie L. Murray, Company L, 9th Infantry, 2nd Division," HI, Box 24242, RG 407, NARA; G. Wilson (1987) *If You Survive* (New York: Ivy Books), p. 31; Stouffer et al., *Studies in Psychology*, Volume II, p. 144. Infantry trainees perceived that medical soldiers other than aid men had less disciplined training periods and directed the derogatory names toward most medical trainees, but once in combat, riflemen called the medics "Doc." One former Battalion Surgeon referred to the aid men as the "real Docs." Questionnaires, "real" Ellis, and Donald F. Eberhart (10 August 2001), and Pomplin.
10. Questionnaire, Johnson.
11. Questionnaires, "infantry man" Stelljes, and Barratt, Fought, Lease, and Manni; Boesch, *Road to Huertgen*, pp. 123–132; Tsuchida *Wear It*, introduction.
12. Questionnaire, John Cramer (12 October 2001).

13 Questionnaire, "irreplaceable" Redfern; B. W. Dohmann (1969), "A Medic in Normandy" *American History Illustrated* (vol. 4, no. 3), p. 12.
14 Questionnaire, Biggins, Bossidy, Irgang, Earl Lovelace (31 August 2001), Manni, Platt, Pomplin, and Sullivan; R. B. Bradley (1970) *Aid Man!* (New York: Robert Bradley), p. 48; "General Orders Number 43, 14 June 1945, After Action Reports, 110th Medical Battalion, 35th Division," Box 9790, RG 407, NARA; Interview, "Marshall O. Miller, 53rd General Hospital, 331st Infantry, 83d Division," HI, Box 2235, RG 407.
15 "General Orders Number 43, 100th Medical Battalion, 14 June 1945," RG 407, NARA.
16 Questionnaires, Barratt, Delaney, DeLoach, Eberhart, Lease, and Pomplin; J. C. McManus (1998) *The Deadly Brotherhood: The American Combat Soldier in World War II* (Navato, CA: Presidio Press), p. 131; Stouffer et al., *Studies, Volume II*, p. 144; Wilson, *If You Survive*, p. 31.
17 Questionnaires, "appeared" Jim Henderson (9 April 2001) and "never saw" Biggins.
18 Questionnaires, Francis Irvin Berry (29 May 2001), Bossidy, and Redfern; Bradley, *Aid Man!*, p. 59.
19 Questionnaires; Wilbur Heinold (2 April 2001) and Winson; Bradley, *Aid Man!*, p. 59.
20 Questionnaire, Maurice DeLoach (20 May 2001).
21 Questionnaires, Berry, Biggins, Delaney, DeLoach, and Platt.
22 The CMB could be awarded retroactively to 7 December 1941. The criteria for award paralleled that of those eligible for the CIB and were intended for medics who accompanied infantry combat troops, engaged the enemy, and worked under fire. "Combat Medical Badge," available from http://www.americal.org/awards/cmb.htm, Internet, accessed 30 July 2002; Questionnaire, Winson.
23 Questionnaires, W. W. Allen (27 November 2001), Carl R. Aschoff (14 June 2001), Stephen J. Barnett (28 June 2001), Brown McDonald, Jr. (22 June 2001), Frank Miller (2 April 2001), Glen J. Mitchell (22 August 2001), Walker Powe (16 August 2001), and Everett Smith (7 August 2001); L. Atwell (1958) *Private* (New York: Simon and Schuster).
24 Questionnaire, McDonald.
25 Questionnaire, Ellis.
26 Ibid.
27 Ibid.
28 Ibid.; Questionnaires, "almost exclusively," McDonald, and Miller.
29 Questionnaire, Pomplin.
30 Questionnaires, "group pressure" Les Habegger (2 April 2001), and Mitchell, Pomplin, Smith, and Donald Warner (16 April 2001).
31 Questionnaires, Mitchell, Smith, Neel Price (5 June 2005), and Warner.

32 Interview, "Battle of Hürtgen Forest, 16 November–3 December 1944, Battalion Aid Men 1st Battalion, 22d Infantry Regiment, 4th Division, with Lieutenant George Kozmetsky, Assistant Surgeon, 1st Battalion, Technical Sergeant, Third Grade Harry I. Fingerroth, 1st Battalion Aid Station, Technical Sergeant, Fifth Grade Joseph J. Thomas, Aid Man with Company B, Technical Sergeant, Fifth Grade Wade H. Carpenter, Aid Man with Company B, by Captain K. W. Hechler, 20 December 1944," CI, Box 24021, RG 407, NARA.
33 Interviews, "Battle of Hürtgen Forest with Kozmetsky, et.al.," NARA.
34 Interviews, "Battle of Hürtgen Forest with Kozmetsky, et.al.," NARA; G. A. Cosmas and A. E. Cowdrey (1992) *The Medical Department: Medical Service in the European Theater of Operations* (Washington, DC: Center of Military History), pp. 365–366.
35 Questionnaire, Jacob E. Way (19 April 2001).
36 Questionnaire, Way.
37 Questionnaire, Smith.
38 Questionnaire, Winson.
39 Ibid.
40 Questionnaires, G. Allen, Bossidy, Fought, and Manni.
41 Questionnaires, Aschoff, Barnett, Berry, Biggins, Braunhardt, Burnett, Andrew Ciampa (7 December 2001), Cramer, DeLoach, Ellis, Fought, Habegger, Heinold, Johnson, Manni, Walter D. Murchison (18 May 2001), Platt, Pomplin, Quint, Redfern, Robert R. Reed II (28 April 2001), Sullivan, and Way; Robert Reed to Dad, 11 February 1945, in Reed's possession; Bradley *Aid Man!*
42 Questionnaire, Braunhardt.
43 Burnett to author, 28 April 2001.
44 Questionnaire, Burnett.
45 R. L. Sanner (1995) *Combat Medic Memoirs: Personal World War II Writings and Pictures* (Clemson, SC: Rennas Productions), p. 95.
46 W. S. Boice (1959) *A History of the Twenty-Second United States Infantry in World War II* (np), p. 28.
47 Bradley, *Aid Man!*, p. 121.
48 "self-doubt," Burnett to author, 7 May 2001; J. Greenberg and H. C. McKeever, eds (1995) *Letters From a World War II GI* (New York: Franklin Watts), p. 46.
49 Richard Quint to author, n.d., in author's possession.
50 Questionnaire, Pomplin.
51 Questionnaire, Johnson.
52 Questionnaires, Barnett, Berry, Ciampa, Fought, Heinold, Irgang, Johnson, Len Karlin (17 July 2001), Pomplin, and Redfern; Quint to author, n.d..
53 Questionnaire, Bossidy.

54 H. P. Leinbaugh and J. D. Campbell (1985) *The Men of Company K: The Autobiography of a World War II Rifle Company* (New York: William Morrow Company), p. 53.
55 Questionnaire, Winson.
56 Questionnaire, Biggins.
57 Ibid.
58 Questionnaires, "would either be killed" Manni, and "the other guy" Winson.
59 Questionnaires, "bringing" Berry, and G. Allen and Johnson.
60 Questionnaires, Lease, Lovelace, and Manni; Bradley, *Aid Man!*, p. 121.
61 Questionnaire, Pomplin.
62 Questionnaires, Barnett, Braunhardt, Heinold, Karlin, Manni, and Sullivan; Bradley, *Aid Man!*, p. 27.
63 W. S. Tsuchida (1947) *Wear it Proudly: Letters* (Berkeley: University of California Press), p. 15.
64 Tsuchida, *Wear It*, pp. 121–123.
65 Questionnaire, Heinold.
66 Robert Reed to mom, 4 March 1945, in Reed's possession.
67 Questionnaire, Allen L. Johnson.
68 Questionnaires, G. Allen, Burnett, Ciampa, Irgang, and Johnson.
69 Questionnaire, Heinold.
70 Questionnaires, Heinold, Miller and Winson; Interview, "Sergeant Joseph Sluzis, 302d Infantry, 94th Division," Box 24240, RG 407, NARA.
71 Questionnaire, Heinold.
72 Ibid.
73 Ibid.

Conclusion

Abstract: *This chapter concludes the book, arguing that while each medical soldier experienced the war uniquely and retained highly personal impressions of the conflict, certain patterns developed in their stories, emerging to give medics a fresh and unique voice in the combat's narrative. With training command focusing on providing inter-changeable soldiers, lessons learned in reference to first echelon care were not apparent until after the war ended. Instead, the medic learned from trial and error and from one another. These first echelon medics who served in Europe, 1944–45, pushed past the inadequacies of their training and creatively adapted to serve best the troops under their care.*

Shilcutt, Tracy. *Infantry Combat Medics in Europe, 1944–45*, Basingstoke: Palgrave Macmillan, 2013.
DOI: 10.1057/9781137347695.

The German invasion of Poland in 1939 shocked Americans long preoccupied with the realities of the Great Depression and lulled by isolationist politicians. Europe's explosion pressed the Roosevelt administration to consider the possibility of a war for which the U. S. Army was woefully unprepared. The 1920s–30s peacetime Army, starved for funds, had operated as a shell, limiting the duties of Medical Department personnel to routine health care needs, and that pattern continued even during mobilization. Although the Medical Department increased in strength after 1939 the primary concern continued to be that of maintaining soldiers' health rather than training for combat demands. By December 1941 medical soldiers could expertly give shots and bandage blistered feet, but they had no practical understanding of caring for battlefield wounded.

Pearl Harbor caught the nation still unprepared, and as the U. S. scrambled to ready the fighting troops, medical training for combat tasks continued drawing little consideration. Most Medical Department training emphasized sterile procedures appropriate for rear echelon medical needs but scarcely related to the coming combat realities the trainees must inevitably confront. Beyond this, the limited field schooling for those medics who trained with the infantry often minimized the role of medics or ignored them altogether. The training which medical soldiers received prepared them for working within a structure that assumed quick and easy recovery of wounded, together with equally quick and easy transport to the rear for definitive care. The spear-point of this system, the company aid man was little considered, grossly underprepared.

War's illusion-shattering power stunned the poorly prepared combat medics swept into the European maelstrom in 1944 and 1945. Any ideas of a paused battle permitting methodical first aid for the combat wounded disintegrated into a kaleidoscopic nightmare of mangled flesh and benumbed minds. Naïve anticipation collided headlong with bitter reality forcing the medical soldiers to confront tasks their training never anticipated. Enlisted men assigned as medics endured a stateside regimen of endless maneuvers, countless miles of daily marches, and rugged bivouacs which hardened them for the soldier's life, but treating blisters and sprains scarcely resembled the mutilation and destruction of combat wounded. Even battalion surgeons schooled in medical institutions found war a surprising, pitiless teacher.

Company aid man Leo Litwak echoed the common experience of countless medics in reflecting on his early enthusiasm for combat:

"I followed my dreams into combat. There I met the dead and dying and faced my own death. It was all I wanted when I first dreamed of war."[1] But a numbing resignation quickly replaced his initial fervor and Litwak's later, mordant feelings resonated with his fellow medics: "we ate among the dead, slept among the dead, tried to rid ourselves of pity for the dead."[2]

The baptism of fire foreshadowed the grim duty to come but combat medics who worked the front lines accepted their life-preserving responsibilities, patching up and moving the wounded men quickly back through the lifesaving system. Aid men enjoyed badly conceived training in treating the wounded, if they had any training at all, yet they understood that their primary care often meant the difference between life and death. So despite the combat environment's initial shock, aid men invariably rallied, adapted to the physical and emotional trauma, learned on the run, and treated their fallen comrades with striking effectiveness. Their adaptability, determination, and ingenuity triumphed over the Army's disheartening neglect of training.

The organizational linch-pin of the medical care system, the BAS, boasted the best-prepared first echelon medical soldiers in the persons of the battalion surgeon and his MAC assistant. And while stateside directives of triage and first aid held generally true on Europe's fighting grounds, the complexities of the physical environment and the surprising severity of the wounds together with the sheer numbers of casualties confounded even those at the BAS. In common with company aid men, these BAS medics initially experienced some disorientation as they entered battle, particularly as they encountered the inadequacies of the medical system. Yet they routinely circumvented textbook protocol to save lives, overcoming initial misconceptions of tending the wounded and working carefully in concert, modifying and fine-tuning procedure in an environment lacking formulaic solutions.

Corpsmen working immediately out of battalion aid stations followed a more traditional military regimen than that of company aid men. Functioning under the direct authority of the battalion surgeon and his designates, these combat medics absorbed the wisdom of veteran detachment members and enjoyed the supportive mutuality of working with like-tasked compatriots. Yet in effectively coordinating the care for combat wounded these medical soldiers also unhesitatingly sidestepped authorized procedure and accepted practices by moving their aid stations forward, assuming responsibilities for which they lacked preparation, and unofficially appropriating equipment or personnel. The presence of the supporting BAS medics boosted the riflemen's confidence level and occasionally BAS personnel developed close relationships with the line soldiers. But BAS activity focused on prompt attention within the stations, readying the wounded for more definitive treatment, and rapid evacuation back through the system, all in support of the initial care given by the company aid men. Together with other front line medics, those in the BAS embraced function over structure, increasing the chances for the wounded soldier's survival.

In sharp contrast to the teamwork of the BAS medics, company aid men worked independently as the initial medical contact within the combat community. Aid men ordinarily labored without direct supervision or chain-of-command accountability as free agents expected to know their jobs and to carry on accordingly. But the realities of limited medical training, the combat line's rapid advance, changing terrain, and weather challenges, all intersected with the necessity of treating multiple casualties to dictate that the aid man adjust constantly. Working in the line of fire, many aid men of course suffered wounds themselves, but often continued their job rather than self-evacuate, driven by the understanding that foot soldiers might not otherwise receive prompt care.[3]

Apart from their combat tasks, company aid men served their combat team by taking on secondary health care responsibilities for which they, again, lacked training. Still the medics adapted by conducting daily sick calls, acting as front line psychiatrists, and providing care and counsel described in no training manual. Although removed from the supportive community of the BAS, and given less pertinent training than any other medical personnel, the company aid man nevertheless functioned effectively and uniquely in the system of care for the wounded soldier.

Infantry linemen, also at that spear's point, gratefully acknowledged the key role medics played and lavished praise on their medical teammates. Riflemen profoundly respected their company aid men, even digging foxholes for them, an honor unlikely accorded to any other soldiers in the European war. At times line soldiers developed close relationships with them, carrying on the fight in better spirit because they knew that "Doc" would always come to their aid. The medic and the infantryman forged a relationship born in blood, yet paradoxically, blood often served to separate them as well.

Removed physically from the BAS community of medical soldiers, company aid men worked among riflemen, even as the non-combat and combat soldiers were performing radically different jobs. Yet beyond these task differentials, the line medics' work intimately and uniquely involved them in their companions' suffering and death. As the conflict continued unrelentingly and casualties mounted, survival instincts forced aid men to raise psychological barriers in order to maintain a bearable distance from the men whom they served. The linemen revered their medics, but the grisly demands of combat aid work, the medics' long hours on the field, and the constant loss of close comrades raised a barrier between the company aid men and their core infantry community, often at great psychological cost. Isolated physically from the BAS and emotionally from their infantry compatriots, company aid men enjoyed no supporting resources.

Autobiographical sketches written immediately following the war as well as accounts published somewhat later reveal medics' emotional struggles as the American army closed on Germany. Perhaps to release their combat demons, these former combat medics crafted powerful word-images laying bare the complex extremes of battle. Their stories juxtapose war's blood and beauty, destruction and delivery, separation and connection, bitterness and innocence. One former 45th Division aid man, frustrated by his effort to convey adequately the intensity of his life in combat's cauldron lamented:

> No one knows what [combat] is like. Our Captain was killed by a sniper. I looked at a soldier shot in [the] head by a sniper. He looked over the wall once too often. Your [sic] in rain, mud over your shoes... I have blood on my clothes and on my hands. The blood dries on [my] hands with the dirt and no water to clean them. Then you eat your rations with this smell on [your] hands. You pray for night because it will be quiet. The rain comes in the foxhole. Your [sic] cold and wet, then you pray for day, so you can dry out.[4]

While some former medical soldiers readily recount their compelling stories, others have found that harrowing memories persist too long after war's end and emotions remain too haunted to allow talk of their war. So they keep silent. But those who do speak out open a darkened window into a grisly task rendered unbearably chaotic by the realities of war.

Notes

1   L. Litwak (2001) *The Medic: Life and Death in the Last Days of World War II* (Chapel Hill: Algonquin Books), p. 3.

2 Litwak, *The Medic*, p. 3.
3 Questionnaires, Ben Burnett (3 July 2001), Warren C. Platt (October 2001), Robert R. Reed II (28 April 2001) and John T. Sullivan (n.d.); G. Wilson (1987) *If You Survive* (New York: Ivy Books), p. 31;; R. B. Bradley (1970) *Aid Man!* (New York: Robert Bradley), p. 64; Interview, "Battle of Hürtgen Forest, 16 November–3 December 1944, Battalion Aid Men 1st Battalion, 22d Infantry Regiment, 4th Division, with Lieutenant George Kozmetsky, Assistant Surgeon, 1st Battalion, Technical Sergeant, Third Grade Harry I. Fingerroth, 1st Battalion Aid Station, Technical Sergeant, Fifth Grade Joseph J. Thomas, Aid Man with Company B, Technical Sergeant, Fifth Grade Wade H. Carpenter, Aid Man with Company B, by Captain K. W. Hechler, 20 December 1944," CI, Box 24021, RG 407, NARA; P. Boesch (1962) *Road to Huertgen: Forest in Hell* (Houston: Gulf Publishing), p. 124; C. B. Currey (1984) *Follow Me and Die: The Destruction of an American Division in World War II* (New York: Stein and Day), p. 107; "General Orders Number 15, 26 February 1945, After Action Reports, 110th Medical Battalion, 35th Division," Box 9790, RG 407, NARA; "General Orders Number 34, 22 May 1945, After Action Reports, 110th Medical Battalion, 35th Division," Box 9790, RG 407, NARA; Interview, "Marshall O. Miller, 53rd General Hospital, 331st Infantry, 83d Division," HI, Box 2235, RG 407, NARA.
4 Questionnaire, G. Allen.

Select Bibliography

Questionnaires

The following respondents answered questionnaires prepared by the author and granted permission for use of the materials. The subjects are all veterans of World War II and the questionnaires are in the author's possession.

Gerald W. Allen
William W. Allen
Jesus (Jesse) Armendariz
Carl R. Aschoff
Stephen J. Barnett
Roy Barratt
Francis Irvin Berry
Walter Biggins
T. William Bossidy
William Braunhardt
Ben Burnett
Andrew Ciampa
Joseph Cosby
John Cramer
Charles C. Cross
Kenneth T. Delaney
Maurice DeLoach
Howard T. East, Jr.
Donald F. Eberhart
Frank R. Ellis
David E. Fought
Les Habegger
James Hanson

Wilbur Heinold
Jim Henderson
Thomas Hoke
Wilfred B. Howsman, Jr.
Frank J. Irgang
Allen L. Johnson
Maurice Kane
Len Karlin
Elvin Keen
Richard L. Lease
Earl Lovelace
Brown McDonald, Jr.
Serge Manni
Frank Miller
Glen J. Mitchell
Walter D. Murchison
Victor Nash
Warren C. Platt
Carroll E. Pomplin
Walker Powe
Neel Price
Richard Quint
Russell Wade Redfern
Robert R. Reed II
John W. Rheney, Jr.
Buster M. Simmons
Everett Smith
Karl Stelljes
John T. Sullivan
Wray Richardson Thomas
Donald Warner
Jacob E. Way
Paul Winson
Angelo Zanin

Correspondence

Ben Burnett to author (March–May 2001), in author's possession.

Sergeant Louis J. Bolla to Charles C. Cross (30 June 1945), in Cross's possession.
Joe Hochadel to author (9 November 2001) in author's possession.
Frank Irgang to author, n.d., in author's possession.
Frank Miller to author (25 April 2001), in author's possession.
Richard Quint to author, n.d., in author's possession.
Robert Reed to parents (1943–45), in Reed's possession.
Karl Stelljes to author (30 May 2001), in author's possession.
Paul Winson to author (11 July 2001), in author's possession.

United States War Department publications

FM 7–30 (1 June 1944) "Service Company and Medical Detachment (Supply and Evacuation) Infantry Regiment." Washington, DC: Government Printing Office.
FM 8–5 (12 January 1942) "Medical Department Units of a Theater of Operations." Washington, DC: Government Printing Office.
FM 8–35 (21 February 1941) "Transportation of the Sick and Wounded." Washington, DC: Government Printing Office.
FM 8–40 (August 1940) "Medical Field Manual. Field Sanitation." Washington, DC: Government Printing Office.
FM 8–45 (October 1940) "Records of Morbidity and Mortality (Sick and Wounded)." Washington, DC: Government Printing Office.
FM 21–11 (7 April 1943) "Basic Field Manual: First Aid for Soldiers." Washington DC: Government Printing Office.
FM 100–10 (December 1940) "Field Service Regulations, Administration." Washington, DC: Government Printing Office.
MTP 8–1 (1942–45) "Mobilization Training Program 8–1." Washington, DC: Government Printing Office.
TM 8–220 (5 March 1941) "War Department Technical Manual: Medical Department Soldier's Handbook." Washington, DC: Government Printing Office.
TO/E 7 series (1941–45) Washington DC: Government Printing Office.
Ginn, L. Homes, Jr. et al. (1945) *Report of the General Board, United States Forces, European Theater Training Status of Medical Units and medical Department Personnel upon arrival in the ETO*. HQ U. S. Forces.

Special reports, interviews, and studies

Interview, "John Spaulding by F. C. Pouge and J. M. Topete," 19 February 1945. Eisenhower Center, University of New Orleans.
Lynch, C., J. Ford, and F. W. Weed (1925) *The Medical Department of the United States Army in the World War*. Washington, DC: GPO.
Record Groups 94 and 407, Records of the Adjutant General's Office 1917–.
Record Group 112, Records of the Office of the Surgeon General (Army).
United States Army in the World War, 1917–1919 (1989–1991). Washington, DC: Center of Military History.

Books and articles

70th Infantry Division Association. "Medic Medic," *The Trailblazer* (Winter 1998): 6–9.
78th Infantry Division Veterans Association. *The Flash* (January 2001): 95.
Amor, J., ed. "Golden Acorn News." *87th Infantry Division Association*. (December 1995): 4.
Ambrose, S. (1997) *Citizen Soldier: The US Army from the Normandy Beaches to the Bulge to the Surrender of Germany*. New York: Simon and Schuster.
Arrington, G. P. (1959) *Infantryman at the Front*. New York: Vantage.
Astor, G. (2000) *The Bloody Forest: Battle for Hürtgen, September 1944–January 1945*. Novato, CA: Presidio Press.
Atwell, L. (1958) *Private*. New York: Simon and Schuster.
Baumer, R. W. and M. J. Reardon. (2004) *American Iliad: The 18th Infantry Regiment in World War II*. Bedford, PA: The Aberjona Press.
Boesch, P. (1962) *Road to Huertgen: Forest in Hell*. Houston: Gulf Publishing.
Boice, W. S. (1959) *A History of the Twenty-Second United States Infantry in World War II*. np.
Bonn, K. E. (1994) *When the Odds Were Even: The Vosges Mountains Campaign, October 1944– January 1945*. Navato, CA: Presidio Press.
Bourke, J. (1999) *An Intimate History of Killing*. New York: Basic Books.
Bradley, R. B. (1970) *Aid Man!* New York: Robert Bradley.
Casey, R. (1945) *This Is Where I Came In*. New York: Bobbs Merrill.

Cassedy, J. H. (1991) *Medicine in America: A Short History*. Baltimore: Johns Hopkins University Press.

Clayman, C. B., ed. (1989) *The American Medical Association Encyclopedia of Medicine*. New York: Random House.

Cotran, R., ed. (1994) *Robbins Pathological Basis of Disease*. Philadelphia: WB Saunders Co.

Chernitsky, D., ed. (1991) *The Foxholes: By the Men of the 110th Infantry, World War II*. Uniontown, PA: Dorothy Chernitsky.

Cooke, E. D. (1946) *All But Me and Thee: Psychiatry at the Foxhole Level*. Washington, DC: Infantry Journal Press.

Cosmas, G. A. and A. E. Cowdrey (1992) *The Medical Department: Medical Service in the European Theater of Operations*. Washington, DC: Center of Military History.

Curley, C. D., Jr. (1998) *How a Ninety-day Wonder Survived the War: The Story of a Rifle Platoon Leader in the Second Indianhead Division During World War II*. Richmond, VA: Ashcroft Enterprises.

Currey, C. B. (1984) *Follow Me and Die: The Destruction of an American Division in World War II*. New York: Stein and Day.

Delaney, J. P. (1946) *The Blue Devils in Italy: A History of the 88th Infantry Division in World War II*. Nashville: The Battery Press.

Dohmann, G. W. "A Medic in Normandy." *American History Illustrated* 4, no. 3 (1969): 8–17.

Doubler, M. (1994) *Closing With the Enemy: How GIs Fought the War in Europe*. Lawrence: University Press of Kansas.

Ellis, J. (1980) *The Sharp End: The Fighting Man in World War II*. New York: Charles Scribner's Sons.

Engle, E. (1967) *Medic: America's Medical Soldiers, Sailors, and Airmen in Peace and War*. New York: John Day Co.

Fishbein, M., ed. (1945) *Doctors at War*. New York: EP Dutton and Co.

Fought, D. "Memories of a 1st Division Medic in World War II." *Bridgehead Sentinel* (Summer 2000): 6.

Franklin, R. "Doc Joe" (2006) *Medic! How I Fought World War II with Morphine, Sulfa, and Iodine Swabs*. Lincoln: University of Nebraska Press.

Fussell, P. (1989) *Wartime: Understanding and Behavior in the Second World War*. New York: Oxford University Press.

Ginn, R. V. N. (1997) *The History of the US Army Medical Service Corps*. Washington, DC: Center of Military History.

Greenberg, J. and H. C. McKeever, eds. (1995) *Letters From a World War II GI*. New York: Franklin Watts.

Greenfield, K. R., R. R. Palmer, and B. I. Wiley. (1947, 2004) *The Organization of Ground Combat Troops.* Washington, DC: Center of Military History, United States Army.

Griesbach, M., ed. (1988) *Combat History of the Eighth Infantry Division in World War II.* Nashville: Battery Press.

Grossman, D. (1995) *On Killing: The Psychological Cost of Learning to Kill in War and Society.* Boston: Little Brown.

Hoegh, L. A. and H. J. Doyle (1946) *Timberwolf Tracks: The History of the 104th Infantry Division, 1942–1945.* Washington, DC: Infantry Journal Press.

Holmes, R. (1985) *Acts of War: The Behavior of Men in Battle.* New York: The Free Press.

Hostetter, P. H. (1999) *Doctor and Soldier in the South Pacific.* Versailles, MO: B-W Graphics. Irgang, F. J. (1949) *Etched in Purple.* Caldwell, OH: The Caxton Printers, Ltd.

Johns, G. S. (1958) *The Clay Pigeons of St. Lô.* Harrisburg, PA: Military Service Publishing Co.

Leinbaugh, H. P. and J. D. Campbell (1985) *The Men of Company K: The Autobiography of a World War II Rifle Company.* New York: William Morrow Company.

Hyman, H. T. (1965) *Handbook of Differential Diagnosis.* London: Pitman Medical Publishing Co., Ltd.

Kemp, H. M. (1990) *The Regiment: Let the Citizens Bear Arms!* Austin: Nortex.

Linderman, G. F. (1997) *The World Within War: America's Combat Experience in World War II.* Cambridge: Harvard University Press.

Litwak, L. (2001) *The Medic: Life and Death in the Last Days of World War II.* Chapel Hill: Algonquin Books.

Lockhart, V. M. (1981) *T-Patch to Victory: The Thirty-Sixth Infantry Division from the Landing in Southern France to the End of World War II.* Canyon, TX: Staked Plains Press.

MacDonald, C. B. (1963) *The European Theater of Operations: The Siegfried Line Campaign.* Washington, DC: Department of the Army.

McBurney, J. B., ed. (1999) *The 13th: A Private's Eye View.* Jennings, LA: John B. McBurney.

McManus, J. C. (1998) *The Deadly Brotherhood: The American Combat Soldier in World War II.* Navato, CA: Presidio.

Mansoor, P. (1999) *The GI Offensive in Europe: The Triumph of American Infantry Divisions, 1941–1945.* Lawrence: University Press of Kansas.

Miller, E. G. (1995) *A Dark and Bloody Ground: The Hürtgen Forest and the Roer River Dams, 1944–1945*. College Station: Texas A&M University Press.

Mittleman, J. B. (1948) *Eight Stars to Victory*. Washington, DC: 9th Infantry Division Association.

Pergrin, D. with E. Hammel (1989) *First Across the Rhine: The 291st Engineer Combat Battalion in France, Belgium, and Germany*. New York: Antheneum.

Phipps, B. (1987) *The Other Side of Time: A Combat Surgeon in World War II*. Boston: Little, Brown and Company.

Pyle, E. (1944) *Brave Men*. New York: Henry Holt.

Sanner, R. L. (1995) *Combat Medic Memoirs: Personal World War II Writings and Pictures*. Clemson, SC: Rennas Productions.

Sewell, P. W., ed. (2001) *Healers in World War II: Oral Histories of Medical Corps Personnel*. Jefferson, NC: McFarland and Co., Inc.

Starr, P. (1984) *The Social Transformation of American Medicine: The Rise of a Sovereign Profession and the Making of a Vast Industry*. New York: Basic Books.

Stouffer, S. et al. (1949) *Studies in Social Psychology in World War II*. Princeton: Princeton University Press.

Taylor, R. R., W. S. Mullins, and R. J. Parks *Medical Training in World War II*. (1974) Washington, DC: US Army.

Towne, A. N. (2000) *Doctor Danger Forward: A World War II Memoir of a Combat Medical Aid Man, First Infantry Division*. Jefferson, NC: McFarland and Co., Inc.

Tsuchida, W. S. (1947) *Wear It Proudly: Letters*. Berkeley: University of California Press.

United States, Army, 1st Division (1995) *First Infantry Division World War II: The Big Red One*. Paducah, Ky.: Turner Pub.

_____, 2nd (1946) *Combat History of the Second Infantry Division*. Nashville, TN: Battery Press.

_____, 78th (1947) *Lightning: The Story of the 78th Infantry Division*. Washington, DC: Infantry Journal Press.

Weigley, R. F. (1981) *Eisenhower's Lieutenants: The Campaign of France and Germany, 1944–1945*. Bloomington: Indiana University Press.

Whiting, C. (1989). *The Battle of Hürtgen Forest*. New York: Orion Books.

Wilson, G. (1987) *If You Survive*. New York: Ivy Books.

Newspapers

Abilene Reporter News. Abilene, Texas, 1941.
Chicago Daily Tribune. Chicago, Illinois, 1943.

Dissertations and theses

Kindsvatter, Peter. "Doughboys, GIs and Grunts: Fear, Resentment and Enthusiasm in the Combat Zone," PhD Dissertation, Temple University, 1978.
Shilcutt, Tracy McGlothlin. "First Link in the Life-chain: Infantry Combat Medics in Europe, 1944–1945," PhD Dissertation, Texas Christian University, 2003.

Electronic resources

"70th Division Association Website." http://www.trailblazersww2.org/amedic.htm, accessed 1 August 2002. Available at http://www.trailblazersww2.org/units_274_accounts_habbeger.htm
"The 100th Division Association." Available http://www.100thww2.org
"Combat Medical Badge." http://www.americal.org/awards/cmb.htm, accessed 10 February 2002.

Index

aid men, see company aid men
American Expeditionary Force (AEF), 5, 13
American Red Cross, 4, 5
Army Medical School, 4
army provisions, 86–7
   boots, 86–7
   coats, 41, 87
   footwear, 41, 87–90
   rations, 51–2, 92–3, 107
   sleeping bags, 86
   socks, 89
   winter gear, 41
Army Specialized Training Program (ASTP), 33
Aschoff, Carl, 20–1, 96n10
assistant battalion surgeons, 21–2, 74n3

BAS medics, 63–81, 121–2
BAS officers, 108
battalion aid station (BAS), 2, 5, 6, 9n13, 52, 63–81, 121
battalion surgeons, 6, 21–4, 41–2, 70, 71, 72, 74n3, 108–9, 121
battlefield medical care, 3–4
Battle of the Bulge, 56, 109
Berry, Francis, 56–7
Biggins, Walter, 107
boots, 87
Bossidy, T. William, 37, 50–1, 58n1, 110
Bradley, Robert, 107–8, 110

Braunhardt, William, 109
Browning Automatic Rifle (BAR), 57
Burnett, Ben, 38–9, 109

Camp Barkeley, 15–16, 21
Camp Butner, 19
Camp Grant, 11, 15, 16, 34, 40
Camp Lee, 15
Camp Reynolds, 12
Camp Robinson, 16
Carlisle Barracks, 20, 22–3
casualties
   access to, 53–4
   finding, 66
   prioritization of, 71
   transport of, 65
chemical warfare, 11, 16, 41
chest wounds, 52–3
citizen-soldiers, 7, 14
Civil War medical care, 4
cold weather disease and injuries, 86–9
collecting companies, 78n26
combat exhaustion/stress, 56, 61n20, 85, 91–4, 102n79, 109, 111, 113
combat infantry badge (CIB), 108
combat initiation experiences, 29–48, 106, 121
combat medics, 2–3, 5–8, 122–3
   see also company aid men
   adjustments made by, 12, 72–3

combat medics – *continued*
  general health care needs treated by, 82–103
  injuries to, 57, 67, 68
  lack of preparation for, 10–24, 29–45, 50–3, 66, 72–4, 84–5, 88, 89, 94–5, 113, 120–1
  noncombatant status of, 67
  relationship with line unit, 18, 19, 35, 104–18, 122–23
  role of, during combat, 49–62
  weapons carried by, 109–10
command post (CP), 71
company aid men, 3, 6, 33–4, 45, 122–3
  see also combat medics
  autonomy of, 109
  combat role of, 49–62
  dissociation of, 109–10
  limits on, 110
  noncombatant status of, 67
  relationship with line unit, 18, 19, 35, 104–18, 122–23
  treatment of health care needs by, 84–95
company aid posts, 5
coping devices, 112
Cosby, Joseph, 57, 62n41
Cross, Charles, 35–7

daily sick call, 84, 94
day-to-day health care, 82–103
D-Day, 31, 32, 34–5, 37, 44, 61n28, 70, 110–11, 113
DeLoach, Maurice, 108
digestive system ailments, 85
disease care, 5

87th Division, 42–3, 83
88th Division, 37, 53, 56–7, 71–2, 89
88th General Hospital, 33–4
Ellis, Frank, 91, 108–9
emergency medical tags (EMTs), 51, 73
emergency supplies, 51, 65
enemy wounded, 83–4
European Theater of Operations (ETO), 12

evacuation systems, 4–6, 12–13, 15, 23, 36, 65–6

fatalism, 93
fear, 92–3, 110
field training, 15, 18–20
1st Division, 32, 88
foot problems, 87–91
footwear, 87–90
4th Division, 61n28, 67, 89, 92
44th Division, 85, 112
45th Division, 41, 123
forward aid stations (FAS), 72–3
Fought, Dave, 32–3
413th Infantry Regiment, 16–17
French Army, World War I, 4
Frenz, Germany, 64
front line troops, relationship between aid man and, 18, 19, 35, 104–18, 122–23
frostbite, 87–8
Fussell, Paul, 43, 96n10

gangrene, 87
German civilians, 84
German prisoners, impressment of, 68
German soldiers, 83–4
Gorgas, William, 4
Gregory, Robert, 2

health care, 82–103
  daily sick call, 84, 94
  digestive system ailments, 85, 94
  foot problems, 87–91
  frostbite, 87–8
  gangrene, 87
  hygiene, lack of, 85, 94
  shoe-pac foot, 89–90
  trench foot, 13, 88–91
  tropical diseases, 4, 41
  weather-related injuries, 13, 56, 86–91
hedgerows, 34–7, 53–5, 57, 65, 70, 73, 110
Heinold, Wilbur, 30, 114–15
Henderson, Jim, 107

hepatitis, 85–6
hospitals, 6–7
hospital training, 19
Hürtgen Forest, 55–6, 68, 89, 92, 109

impressment, 68–9
Indianhead Division, 34
Irgang, Frank, 33–4, 110
isolation, emotional, 36, 45, 106, 110, 113–14
Italy, 37, 42, 44, 53, 56–7, 71, 89

Johnson, Allen, 34–5, 54, 107, 113

Lawson General Hospital, 11–12
Lease, Richard, 42–3
litter bearers, 2, 3, 6, 11, 13, 17, 19, 23, 32–3, 36–8, 45, 50, 52, 56, 64–74, 74n3, 78n26, 80n46, 90, 95n7, 109
Litwak, Leo, 120–1
Lovelace, Earl, 44–5, 57, 87–8

MAC officers, 79n29
malingering, 14
Martinez, Ben, 37
McDonald, Brown, 24, 41–2, 89, 92–3
Medical Administrative Corps (MAC), 3, 19–22
Medical Administrative Corps-Officer Candidate School (MAC-OCS), 20–2
Medical Battalion, 6
Medical Corps (MC) officers, 3
Medical Department
  Civil War, 4
  historical development of, 3–5
  lack of preparedness of, 120
  marginalization of, 14
  medical soldiers in, 7, 10–28
  shortages, 4–5, 13, 14, 16
  World War I, 4, 12–14
medical doctors, 22–4, 41–2, 74n3
medical equipment, 53, 86
Medical Field Service School, 20
medical helmet markings, 67–8
medical kits, 51, 52

Medical Replacement Training Center (MRTC), 11, 15–17
medical techniques, 52–3
medics, *see* BAS medics; combat medics; company aid men
military history, 7–8
military medicine, 4, 6–7, 13–14
Miller, Frank, 16–17, 64
mine detectors, 55
Mitchell, Glen, 83
Mobilization Training Program, 14, 16
morale, 107–8
motorized evacuation, 69
multiple casualties, 53

neuro-psychiatric casualties, 91–4

officer candidate schools, 20
100th Division, 40–2, 68, 110
104th Division, 38, 39, 110

peacetime training, 14
Pershing, John J., 14
physical exhaustion, 93
Pomplin, Carroll, 110, 112
provisions, *see* army provisions
psychiatric illnesses, 91–4
psychological ailments, 14
Pyle, Ernie, 32

rear echelon, 69, 78n26
Red Cross insignia, 67–8
Redfern, Russell, 53
Reed, Robert, 19–20, 112
Reed, Walter, 4
regimental surgeon, 73
rehabilitation centers, 6–7

Sanitary Commission, 4
School for Battalion Surgeon's Assistants (SBSA), 21–2
2nd Division, 34, 35, 44, 40, 51, 54, 87
self-inflicted wounds (SIW), 93
70th Division, 69
78th Division, 90

sexually transmitted diseases (STDs), 16–17, 96n10
shell fright, 14
shoe-pac foot, 89–90
Siegfried Line, 39, 50, 87–8
63rd Division, 65
Smith, Everett, 67, 109
St. Lô, 37, 66
Sullivan, John, 53
supplies, medical, 51, 65
Surgeon General, 3
surrenders, by enemy soldiers, 83

30th Infantry Division, 12, 39, 40, 53, 54, 66, 67, 89, 105
31st Infantry Division, 18
35th Infantry Division, 68, 107
311th Infantry Regiment, 27n34
tourniquets, 53
Towne, Allen, 95n7
training
    field, 15, 18–20
    hospital, 19
    inadequacies of, 10–24, 29–45, 50–3, 66, 72–4, 84–5, 88, 89, 94–5, 113, 120–1

MAC, 19–22
mobilization, 14, 16
peacetime, 14
stateside, 10–28
wartime, 14
    during WWI, 12–14
trench foot, 13, 88–91
tropical diseases, 4, 41
Tsuchida, Frank, 112
22nd Infantry Regiment, 75n5, 101n62, 117n32, 124n3
28th Division, 61n28, 89

venereal diseases (VD), see sexually transmitted diseases (STDs)
Vosges Mountains, 24, 40–2, 44, 65

walking wounded, 52
war neuroses, 14
Way, Jacob, Jr., 109
weather-related injuries, 13, 56, 86–91
Winson, Paul, 11–12, 39–40, 105–6, 109
winter conditions, 56, 86–9
Wood, Leonard, 4
World War I, 4, 12–14
World War II, 2–3, 5–7

The manufacturer's authorised representative in the EU is Springer Nature Customer Service Centre GmbH, Europaplatz 3, 69115 Heidelberg, Germany. If you have any concerns regarding our products, please contact ProductSafety@springernature.com

Printed and bound by CPI Group (UK) Ltd, Croydon, CR0 4YY

23/03/2026

02076355-0017